Biochemistry Research Trends Series

AF849429

GLYCOLYSIS: REGULATION, PROCESSES AND DISEASES

Biochemistry Research Trends Series

Glycolysis: Regulation, Processes and Diseases
Paul N. Lithaw (Editor)
2009. ISBN: 978-1-60741-103-1

Biochemistry Research Trends Series

Glycolysis: Regulation, Processes and Diseases

Paul N. Lithaw
Editor

Nova Biomedical Books
New York

Copyright © 2009 by Nova Science Publishers, Inc.

All rights reserved. No part of this book may be reproduced, stored in a retrieval system or transmitted in any form or by any means: electronic, electrostatic, magnetic, tape, mechanical photocopying, recording or otherwise without the written permission of the Publisher.

For permission to use material from this book please contact us:
Telephone 631-231-7269; Fax 631-231-8175
Web Site: http://www.novapublishers.com

NOTICE TO THE READER

The Publisher has taken reasonable care in the preparation of this book, but makes no expressed or implied warranty of any kind and assumes no responsibility for any errors or omissions. No liability is assumed for incidental or consequential damages in connection with or arising out of information contained in this book. The Publisher shall not be liable for any special, consequential, or exemplary damages resulting, in whole or in part, from the readers' use of, or reliance upon, this material. Any parts of this book based on government reports are so indicated and copyright is claimed for those parts to the extent applicable to compilations of such works.

Independent verification should be sought for any data, advice or recommendations contained in this book. In addition, no responsibility is assumed by the publisher for any injury and/or damage to persons or property arising from any methods, products, instructions, ideas or otherwise contained in this publication.

This publication is designed to provide accurate and authoritative information with regard to the subject matter covered herein. It is sold with the clear understanding that the Publisher is not engaged in rendering legal or any other professional services. If legal or any other expert assistance is required, the services of a competent person should be sought. FROM A DECLARATION OF PARTICIPANTS JOINTLY ADOPTED BY A COMMITTEE OF THE AMERICAN BAR ASSOCIATION AND A COMMITTEE OF PUBLISHERS.

Library of Congress Cataloging-in-Publication Data

Glycolysis : regulation, processes, and diseases / editor, Paul N. Lithaw.
p. ; cm. -- (Biochemistry research trends)
Includes bibliographical references and index.
ISBN 978-1-60741-103-1 (hardcover : alk. paper)
1. Glycolysis. I. Lithaw, Paul N. II. Series: Biochemistry research trends.
[DNLM: 1. Glycolysis--physiology. QU 75 G56803 2009]
QP701.G58 2009
572'.567--dc22

2009004641

Published by Nova Science Publishers, Inc. ✝ New York

Contents

Preface

Glycolysis literally means "splitting sugars." In glycolysis, glucose (a six carbon sugar) is split into two molecules of a three-carbon sugar. Glycolysis yields two molecules of ATP (free energy containing molecule), two molecules of pyruvic acid and two "high energy" electron carrying molecules of NADH. Glycolysis can occur with or without oxygen. In the presence of oxygen, glycolysis is the first stage of cellular respiration. Without oxygen, glycolysis allows cells to make small amounts of ATP. This process is called fermentation. This new book presents the latest research in the field.

Chapter I - The lactic acid bacterium *Lactococcus lactis* has been exploited for centuries in the production of fermented foods. Through the homofermentative conversion of sugar to lactate, the resulting acidification preserves the fermented food while it contributes to the development of desired texture and organoleptic qualities. As an industrial microorganism, *L. lactis* is used, apart from food fermentations, in the production of lactic acid and the bacteriocin nisin. Both of them are products of high value and of extensive use in the food industry. The low specific productivity obtained in the most successful fermentation systems, a characteristic of both *L. lactis* products, is the major cost increasing factor and at the same time the factor that triggers research in the areas of sugar transport, glycolysis and the shift between the homofermentative and heterofermentative metabolism. From an industrial point of view there is much interest to increase the overall flux through glycolysis and to control the production of other end products than the desired. The regulation of glycolysis and the shift between different fermentation pathways have been extensively studied. The mapping however, of regulatory mechanisms does not necessarily lead to an understanding of which enzymes have control on the flux. Today, despite the wealth of metabolic information collected during years of intensive research and numerous genetic tools available for *L. lactis*, the fundamental question of what controls the glycolytic flux in this organism still represents a black box.

Chapter II - Everyone of true conscience must admit that over the last 30 years insufficient progress has been made in the "war to cure cancer." Otto Warburg, M.D., Ph.D., showed decades ago that development of cancer had a singular, prime cause. Each and every time cells (and tissues) were deprived of oxygen for a sufficient period of time, cancer developed. Furthermore, he clearly showed that the distinguishing feature of all cancer cells is the increase of anaerobic glycolysis and concurrent decrease of respiration—not merely excessive cell divisions. The significant increase in glycolysis observed in tumors has been verified today, yet few oncologists or cancer researchers understand the full scope of Warburg's work and its great importance. Without the use of Warburg's seminal discovery, cancer can never be truly cured—merely treated—although ineffectively, because when cancer returns from "remission," as is often the case, the patient has a high probability of death; treatments are ineffective. Extensive references to Warburg's original research are given.

Chapter III - The effect of complexing reaction of the activator ADP has been investigated in a model glycolytic reaction-diffusion system generating Hopf and Turing wave instabilities. The complex formation reaction with the activator species reduces drastically the domain of homogeneous Hopf bifurcation in the parameter space producing more Turing region, where Turing bifurcation may be initiated by inducing inhomogeneous perturbations. Our numerical results conform to the expectation that Hopf wavelengths depend strongly on the degree of complexing reaction of the activator, whereas Turing wavelengths don't. For this model system, the reaction velocity and entropy production as a function of the reaction affinity are computed and the results interpreted in terms of the efficiency of biochemical engines.

Chapter IV - For most human at least 50 % of the dietary energy comes from carbohydrates. Skeletal muscles make up 30-40 % of the body weight and the major part of the carbohydrate is stored as muscle glycogen ($\approx$ 80 %). After a carbohydrate meal $\approx$ 35 % of the carbohydrates are stored as muscle glycogen whereas 20 % ends up as liver glycogen. A major part of ingested carbohydrates, therefore, passes through glycolysis in skeletal muscles. Glycolysis in skeletal muscles is activated during insulin-mediated glucose disposal and more in muscles with high glycogen content. Skeletal muscles cannot release glucose molecules because glucose 6-phosphatase is lacking. However, muscle glycogen can be metabolised via glycolysis and released as lactate; skeletal muscles are the major contributor of blood lactate appearance. The released lactate is the major substrate for gluconeogenesis or oxidation in some tissues. Adrenaline-mediated glycogen phosphorylase activation initiates glycolysis in resting skeletal muscles. Skeletal muscle glycolytic rate is highest during high intensity exercise when muscles convert chemical energy to movement. During exercise, glycolytic rate in skeletal muscles can increase more than 100-fold, and substantial amount of glycogen is rapidly broken down in muscles. In the present paper, regulation of glycolysis in skeletal muscles by insulin, adrenaline and exercise is discussed. Furthermore, the physiological role of skeletal muscles glycolysis for whole body metabolic regulation in normal and type 2 diabetes is addressed.

Chapter V - The lung is an organ with respiratory and non-respiratory functions. As such it plays a critical role in maintaining homeostasis in the body. Various cell types occur which play a role in maintaining lung structure and function. Glucose is a major energy substrate and also plays a central role in lung development. Certain cells in the lung, for example the type I pneumocytes depends largely on glycolysis for energy. Most of the glucose used by the lung is converted to lactate. The flux of glucose through the glycolytic pathway is controlled.

Apart from its role in energy metabolism, glycolysis also plays and important role in apoptosis in the lung. In addition to playing an important role in the flow of glucose through the glycolytic pathway, evidence suggests that glyceraldehyde-3-phosphate dehydrogenase, also plays a role in induction of apoptosis. In addition it also serves as an intracellular sensor for oxidative stress that may play an important early role in the cascade of reactions leading to apoptosis.

Glycolysis is also necessary for normal aging of lung cells and thus the maintaenance of lung structure and function. It has been shown that suppression of glycolysis induces premature aging in lung. This adversely affects maintenance of lung structure and function and increased susceptibility to respiratory diseases.

A number of studies showed that maternal nicotine exposure during gestation and lactation resulted in an irreversible inhibition of glycolysis. The site of action appears to be at the phosphofructokinase level. It is proposed that this inhibition is due to a change in the program that controls glucose flux though glycolysis by oxidant effects of nicotine. It is also suggested that the permanent inhibition may probably result in premature aging of the lungs of the offspring that was exposed to nicotine via the placenta and mother's milk. This means that using nicotine replacement therapy to quit smoking during gestation and lactation is not advisable.

Chapter VI - *Escherichia coli, Bacillus subtilis* and *Saccharomyces cerevisiae* are well-characterized species which have contributed significantly to our present knowledge of central metabolism. In addition to their roles as model organisms in biology, they are also widely used as microbial cell factories for the biotechnological production of valuable products like insulin and vaccines. Glycolysis is the core pathway for carbon metabolism in these cells to provide the necessary energy and carbon backbones for product synthesis and cellular growth. Carbon fluxes through glycolysis have evolved to be under rigid regulatory control so as to coordinate catabolic fluxes with biosynthetic demands during growth. While the control of activity of glycolytic enzymes through allosteric regulation is well-understood, the regulation of glycolytic genes at the transcriptional level has begun to attract attention only recently. Additionally, a few post-transcriptional regulators were also found to regulate glycolysis at the level of mRNA stability. This communication will describe our current knowledge on glycolysis-related transcriptional/post-transcriptional factors regulating mRNA synthesis and degradation in these three representative microbial cell systems.

Chapter VII - The purpose of this study was to analyse values generated during 30 s of high intensity cycle ergometry exercise when cradle resistive forces were calculated from total - body mass (TBM) or fat - free mass (FFM). A further aim was to compare the power values generated with performance indices recorded during maximal running performance on a modified multi stage fitness test and to validate the running test as a measure of anaerobic performance. Body density was calculated using underwater weighing procedures. Fat mass was estimated from body density values. Significant differences ($P < 0.01$) were observed between the TBM and FFM protocols for peak power output (PPO; 1264 ± 156W vs. 1366 ± 177W respectively). Significant differences ($P < 0.01$) were also recorded between the TBM and FFM protocols for resistive force selection and pedal revolutions (7.3 ± 1.2 vs. 6.2 ± 1.1 kg; 136 ± 8.7 vs. 144 ± 7.6 rpm respectively). There were no differences ($P > 0.05$) recorded between mean power output (MPO) or fatigue index (FI %). Values recorded for the running test were 71.4 ± 7.5 s. Significant ($P < 0.01$) linear relationships were found between PPO and running times for both the TBM and FFM protocols with more of the variance accounted for during the FFM protocol. Blood lactate concentrations increased significantly from rest to

5 min post exercise for all three experimental conditions and were highly correlated ($P < 0.01$). Results from the study suggest that higher PPO values are obtainable when resistive forces used in high intensity cycle ergometry exercise reflect lean tissue mass. Also, the running test proved to be a viable measure for the quantification of high intensity running performance during periods of intense work.

Chapter VIII - ***Background:*** Energy system interaction during repeated bouts of maximal activity is complex, and relatively little is known about different energy system contribution during exercise.

Objective: The aim of the present study was to examine the contribution of anaerobic glycolysis to a repeat sprint protocol via the assessment of blood lactate concentration.

Research design and methods: Eight male, healthy subjects volunteered to participate in the study. The subjects performed eight 6-s sprints on a friction loaded cycle ergometer with a 60-s recovery period between each sprint. Plasma volume corrected blood samples were collected at rest (following 30 min in a supine position), following each sprint (within the first 10-s) and at 5 min post-exercise.

Results: The highest mean (MPO) and peak power output (PPO) was observed in the first and third sprint for both conditions (777.3 ± 142.2 W and 874.9 ± 175.6 W, respectively; see figure 1). Power outputs were maintained during the exercise period with no significant decreases observed between sprint 1 and eight ($P > 0.05$). In contrast, blood lactate concentrations increased throughout the successive sprint periods from a resting value of 0.67 ± 0.47 mmol/L, to a peak value of 7.5 ± 1.8 mmol/L, immediately following sprint 8 ($P < 0.05$) Plasma volume changes showed a gradual haemoconcentration after sprint two (-0.86 ± 5.94%), and approached a significant change from the resting value immediately after sprint eight (~9.5% haemoconcentration; $P < 0.05$).

Conclusions: The main findings of this study were that 60-s recovery from brief maximal exercise is sufficient to replenish the anaerobic energy stores (ATP-PC).and that anaerobic glycolysis plays a significant role in energy provision as exercise progressed.

Chapter IX - Glycolysis is probably the best understood biochemical pathway. It has been subjected to about every imaginable type of investigation, from phenomenological observations to detailed analyses of its components with methods of enzyme kinetics and *in vivo* nuclear magnetic resonance. In many ways, the gradual increase in information and knowledge associated with the glycolytic pathway can be seen as a representative of the growing body of insights into metabolism in general. The development of mathematical and computational models of glycolysis has mirrored the experimental exploration, although with substantial time delay. Indeed, models of glycolysis can be seen as sentinels of important phases of metabolic model creation, including the choices of model types at different times and the purposes for creating these models. Early models were designed as proof of concept that mathematical equations were capable of capturing biological observations. Some of these early, simple models eventually grew into comprehensive descriptions of glycolysis in different contexts and with species dependent variations, allowing detailed simulations of what-if scenarios. Other models stayed intentionally simple in order to allow the extraction and rigorous mathematical analysis of the essence of the pathway, for instance, with respect to oscillations. Some of the models were used for optimization within a context of metabolic engineering, others as means of explaining non-intuitive features of pathway control and regulation. This chapter reviews some of these developments and demonstrates how they have been leading to the present-day frontiers of discovering design and operating principles and to guidance for the creation of pathway systems in the new field of synthetic biology.

Chapter X - Induced hypothermia after cardiopulmonary resuscitation ameliorates neurological outcome and is currently considered standard of care in clinical practice. An increasing amount of reports indicate that induced hypothermia is also beneficial in other conditions of hypoxia–induced organ injury, including intestinal ischemia–reperfusion injury and acute lung injury. Hydrogen sulphide, which inhibits oxidative phosphorylation, has been used to induce a suspended animation–like state in several rodent models, resulting in hypothermia and a reduction in metabolic rate. Hydrogen sulphide has been found to be protective against ischemia–reperfusion induced organ injury, including gut ischemia and acute lung injury.

In this manuscript, we speculate on the potential therapeutic effects of reducing metabolism in critically ill patients. In these patients, an exaggerated inflammatory response is common, which often results in multiple organ injury. Inducing a hypometabolic state during critical illness may limit organ injury by reducing oxygen consumption, a novel approach in the treatment of critically ill patients. Mitochondrial dysfunction during critical illness is described and the potential therapeutic possibilities of influencing metabolism during critical illness is discussed. Methods of inducing hypothermia and of inducing a suspended animation–like state with the use of hydrogen sulphide are described.

Short Communication - Enhanced glycolysis is observed in most of cancerous cells and tissues, called as the Warburg effect. The clinical significance of the Warburg effect has been well established, while it is not completely clarified why and when cancer cells start to display and acquire such a characteristic metabolic property. Especially cancerous cells maintain enhanced glycolysis in tissue culture under standard condition (20% oxygen), which can not be explained by the cellular adapataion to hypoxic condtion via transcriptional factor HIF-1 (Hypoxia inducible factor-1) activation. Recent findings on senescent and cancer research discovered the unexpected role of the Warburg effect in protecting cells from oxidative damage. These anti-ageing effect of the Warburg effect can be a clue to understand pathophysiological impact of such metabolic shift in tumorigenesis.

In: Glycolysis: Regulation, Processes and Diseases
Editor: Paul N. Lithaw
ISBN: 978-1-60741-103-1
© 2009 Nova Science Publishers, Inc.

Chapter I

Regulation of Glycolysis in *Lactococcus Lactis*

Maria Papagianni
Department of Hygiene and Technology of Food of Animal Origin,
School of Veterinary Medicine, Aristotle University of Thessaloniki,
Thessaloniki, 54006, Greece

The lactic acid bacterium *Lactococcus lactis* has been exploited for centuries in the production of fermented foods. Through the homofermentative conversion of sugar to lactate, the resulting acidification preserves the fermented food while it contributes to the development of desired texture and organoleptic qualities. As an industrial microorganism, *L. lactis* is used, apart from food fermentations, in the production of lactic acid and the bacteriocin nisin. Both of them are products of high value and of extensive use in the food industry. The low specific productivity obtained in the most successful fermentation systems, a characteristic of both *L. lactis* products, is the major cost increasing factor and at the same time the factor that triggers research in the areas of sugar transport, glycolysis and the shift between the homofermentative and heterofermentative metabolism. From an industrial point of view there is much interest to increase the overall flux through glycolysis and to control the production of other end products than the desired. The regulation of glycolysis and the shift between different fermentation pathways have been extensively studied. The mapping however, of regulatory mechanisms does not necessarily lead to an understanding of which enzymes have control on the flux. Today, despite the wealth of metabolic information collected during years of intensive research and numerous genetic tools available for *L. lactis*, the fundamental question of what controls the glycolytic flux in this organism still represents a black box.

Introduction

Since the early years of the last century, there has been an enormous increase in the industrial production of cheeses and fermented dairy products. Today, the dairy industry represents the most dynamic among the food industries, with a high level of mechanization and large factory sizes that process increasing quantities of milk daily aiming at shorter processing times. This is reflected in enormous demands at the starter cultures that must be of known and controlled activity, stable quality and resistance to bacteriophages.

Lactic acid bacteria (LAB) are used widely in the production of fermented food products due to their specific metabolic activities, which translate into technological, nutritional and health properties. These metabolic activities result in the production of lactic and other organic acids, polysaccharides, various volatiles, and effective antimicrobials, known as "bacteriocins", many of which are established industrial products of microbial origin. LAB industrial fermentations are carried out for the production of pure or mixed cultures, lactic acid, polylactic acid, polysaccharides, the bacteriocin nisin, and a plethora of fermented food products. The fermented dairy products represent a large market share of dairy products, and increase developments of products containing nutraceutical cultures are anticipated for North American, European and Japanese markets. Lactic acid bacteria are already used in many probiotic dairy products marketed worldwide. Therefore, the large demand underlines the economic importance of the large-scale applications of LAB. Most of LAB fermentations are characterized by low specific productivities and low specific glucose uptake rates, facts that translate into products of high value. In view of the economic importance of LAB fermentations produce, sugars metabolism, and in particular lactose metabolism, has been the subject of considerable research aiming at understanding, and more recently, exploiting the process involved.

Since the 1980s a large number of research articles and reviews, many of which have been of outstanding quality, have been published on the metabolism, physiology, genetics, fermentation, production and use of industrial lactic acid bacteria, as well as on the genetics, production and applications of the bacteriocins produced by lactic acid bacteria. Reviewing the literature, one can point out a central issue, from several points of view, in the catabolism of sugars by LAB. The history of the genetics of lactose utilization stretches back to the 1930s when researchers observed the loss of lactose metabolism in *Lactococcus lactis* [1], but only during the early 1970s this was explained by the plasmid-located nature of the lactose genes [2]. This signaled the way for a detailed genetic analysis of the *lac* genes that has produced the first model for gene organization and regulation in *L. lactis* [1, 3]. Subsequently, the genetics of lactose, and the related sugar galactose, metabolism has been the subject of several reviews [4, 5]. Parallel with the genetic analysis, has been the progress on metabolic analysis. The phosphoenolopyruvate – phosphotransferase system (PEP-PTS) was detected and described in *L. lactis* in 1969-1970 [6, 7]. Later, in 1978, Thompson [8] reported on the *in vivo* regulation of glycolysis and characterization of sugar:phosphotransferase systems in *L. lactis* and identified a key role of the allosteric enzyme pyruvate kinase in the regulation of glycolysis and phosphotransferase system. During the 1980s, Thompson and co-workers [9, 10, 11, 12] published in depth investigations on sugar uptake and metabolism and in particular the regulation of glycolysis in *L. lactis*.

Since then, a significant number of important works on the metabolism of sugars and glycolysis have been published, mainly with *L. lactis*. The wealth of information on genetics and metabolism of this bacterium, the availability of genetic tools and the complete genome sequence [13] consolidated its status as a model LAB.

The aim of this chapter is to present an overview of the progress made in research on the regulation of glycolysis in, by far the most extensively studied lactic acid bacterium, *Lactococcus lactis*. Genes, sugar transport mechanisms, enzymes involved, and pools of metabolites, will be discussed. The impact and the potential of genomics on the study of the regulation of glycolysis will be demonstrated by surveying the published research of the last 30 years.

Sugar Uptake and Initial Metabolism in *Lactococcus Lactis*

The first step in the metabolism of sugars is their transport across the cell membrane. Carbohydrate transport in bacteria can be achieved by three major uptake systems: 1) the PTS, phosphoenolopyruvate:phosphotransferase system, in which, apart from transport, phosphorylation of sugar takes place [14]; 2) ion-linked transport [15]; and 3) ABC transport systems, which are primary transport systems that couple ATP hydrolysis with translocation [16]. *L. lactis* typical industrial strains transport lactose into the cell via a highly efficient phosphoenolopyruvate:phosphotransferase system (PEP:PTS) with concomitant phosphorylation of sugar.

From an energetic point of view, the PEP:PTS is probably the most efficient sugar transport process since the sugar is translocated and phosphorylated in a single step at the expense of one PEP molecule. This is equivalent to one ATP molecule, since in the glycolysis one ATP molecule is derived from one PEP molecule at the level of pyruvate kinase reaction. For sugars that are accumulated by other transport systems, more ATP molecules are required for transport and phosphorylation.

The PEP:PTS in Lactococci

The structure and function of the enzymes involved in the lactose PEP:PTS have been reviewed extensively [1, 14, 17]. The PTS is a group translocation process in which the transfer of the phosphate moiety of PEP to carbohydrates is catalyzed by the general non-sugar specific proteins, the enzyme I (EI) and the heat stable protein (HPr), in combination with the sugar specific enzyme II (EII) proteins. Following autophosphorylation of enzyme I at the expense of PEP, enzyme I catalyzes the phosphorylation of HPr at histidine 15, resulting in HPr(His-P). The phosphate group from this complex is then transferred to the sugar substrate by a specific enzyme II that transfers and phosphorylates the sugar. The internalized disaccharide is hydrolyzed by the phospho-β-galactosidase into galactose-6-phosphate and glucose. Glucose is then metabolized in the reactions of the tagatose-6-phosphate pathway into triose-phosphate (figure 1) [1].

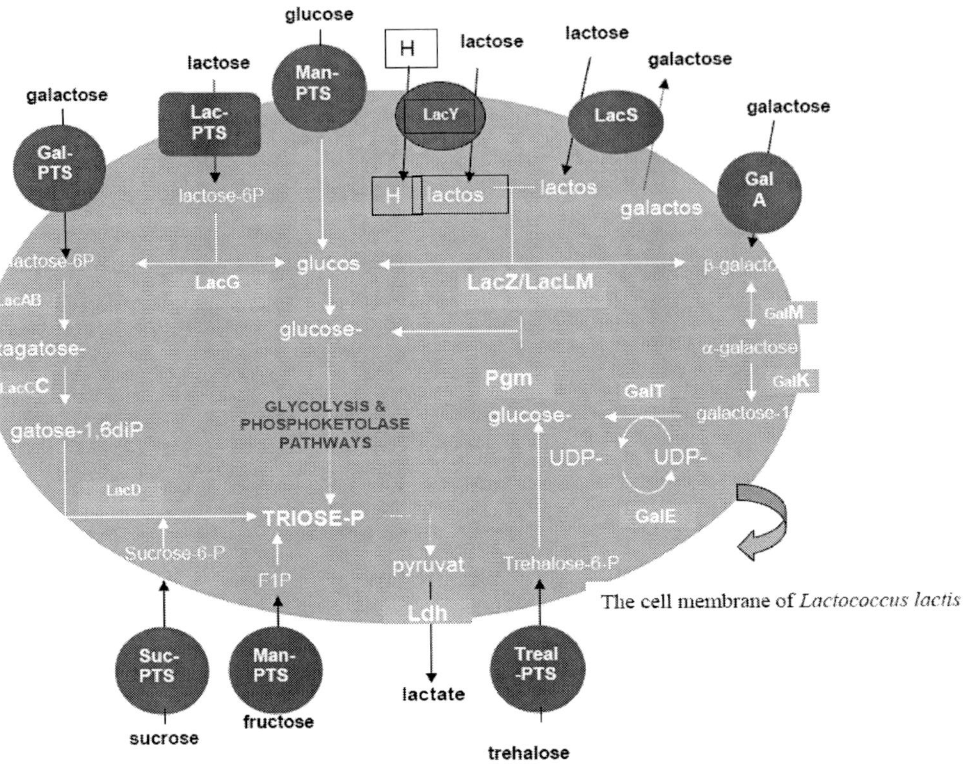

Figure 1. Metabolic pathways involved in carbohydrate metabolism in *Lactococcus lactis.*

EII proteins may consist of one or more proteins and are composed of three domains: the EIIA and EIIB domains, involved in phosphotransfer, and the membrane located EIIC domain, which is most likely involved in translocation of the sugar substrate. When two or more EII proteins are involved, one is always membrane bound (e.g., EIIC), while the other one is soluble (e.g., EIIA) [18].

The genes encoding HPr and EI, *ptsH* and *ptsI*, respectively, have been cloned in several bacteria and *L. lactis* and found often to be organized in an operon structure with the gene order *ptsHI* [19].The *L. lactis ptsH* and *ptsI* genes, encoding the general proteins of the phosphoenolopyruvate-dependent phosphotransferase system, HPr and enzyme I, respectively, were cloned and the regulatory role of HPr was studied by mutation analysis of its gene by Luesink et al. [19]. The *ptsH* gene was transcribed as a single 0.3 kb mRNA but also as a part of a longer 2.0 kb mRNA with he *ptsI* gene. Expression of the operon was regulated at the transcriptional level and glucose-inducible but the regulatory elements have not yet identified. Disruption of the *ptsH* and *ptsI* genes, in *L. lactis* NZ9800, resulted in a reduced growth rate at the expense of glucose, but no growth at the expense of fructose and sucrose, confirming the dominant role of the phosphotransferase system in the uptake of these sugars in *L. lactis* and also the presence of another, non-PTS, transport system for glucose.

Apart from its function in the uptake of sugars, the PTS also plays a regulatory role described in both Gram-positive and Gram-negative bacteria [14]. In Gram-negative bacteria the PTS regulates the concentration of cAMP via activation of adenylate cyclase by the phosphorylated form of glucose-specific EIIA, the concentration of which increases in the absence of PTS substrates. Elevated cAMP concentrations lead to transcriptional activation of several genes via the binding of the cAMP receptor protein complexed with cAMP to operator sites located in the promoter regions of affected genes. Furthermore, the unphosphorylated form of the glucose-specific EIIA reduces the uptake of several non-PTS sugars via an interaction with the uptake protein.

In Gram-positive bacteria, the HPr(His-P)-mediated phosphorylation of two glycerol kinases results in an increased activity of both enzymes in *Enterococcus* spp. [20, 21] In contrast, enzyme I/HPr(His-P)-mediated phosphorylation of the lactose permease in *Streptococcus thermophilus* results in a reduced permease activity leading to a decreased uptake of sugar [22].

Apart from phosphorylation at residue His-15, a second phosphorylation site has been identified in HPr, the function of which has been shown only in Gram-positive bacteria [23]. Phosphorylation at Ser-46 is catalyzed by an ATP-dependent protein kinase that is activated by fructose-1,6-bisphosphate [24, 25]. The genes encoding the two enzymes involved, have been cloned and their involvement in the phosphorylation of HPr at Ser-46 has been established [23]. This seryl-phosphorylated form of HPr, designated as HPr(Ser-P), 1) interacts with several PTS and non-PTS sugar permeases, the process termed inducer exclusion and results in reduced sugar uptake rates; 2) it allosterically activates sugar-phosphate phosphatases in *L. lactis* (and others) that catalyze the dephosphorylation of various phosphorylated sugars, resulting in an efflux of the sugar from the cell, a process known as sugar expulsion; and 3) it can negatively regulate the transcription of genes by an interaction with the catabolite control protein CcpA [1, 14, 23].

The participation of HPr(Ser-P) in the CcpA-mediated transcriptional activation of the *las* operon in *L. lactis* has been shown by Luesink et al. [19, 26]. Growth on glucose resulted in higher activities of the glycolytic key enzymes phosphofructokinase (PFK), pyruvate kinase (PYK), and the L-lactate-dehydrogenase (LDH), the genes of which form the tricistronic *las* operon. This indicated that CcpA might act as a transcriptional activator. However, deletion of the *ptsH* gene led only to a 30% reduction of the glycolytic enzyme activities, indicating the regulation of the *las* operon is not exclusively dependent on an intact *ptsH* gene. Thus, it is possible that other effectors of CcpA are involved in *las* operon activation.

Genetics of the Lactose-PTS in Lactococci

Gasson and co-workers [27, 28] first showed that the genes encoding the PEP:PTS and the tagatose-6-phosphate pathway in a L. lactis strain are plasmid-located. These are the genes lacE and lacF, encoding EIIBC and EIIA, the lacG, encoding phospho-β-galactosidase, the lacAB, lacC and lacD, encoding the tagatose-6-phosphate enzymes galactose-6-phosphate isomerase, tagatose-6-phosphate kinase and the tagatose 1,6-diphosphate aldolase [29, 30, 31, 32]. The genes are organized in the order of lacABCDFEGX in a 7.8 kb operon. Next to lacX an iso-ISS1 element was identified. The transcriptional regulator LacR of the operon is positioned upstream and in an orientation towards the operon so that the two promoters are in back-to-back configuration [31].

The lac genes are transcribed as two transcripts, the 6 kb lacABCDFE and the 8 kb lacABCDFEGX genes. The lacX gene has been shown to be dispensable for growth on lactose [1]. The whole lac operon is induced up to 10-fold with growth on lactose. The lacR promoter is induced during growth on glucose [1]. LacR belongs to the family of DeoR repressors and is responsible for both repression and activation of the lac operon [31, 33]. This is achieved through a high affinity operator, lacO1, and a lower affinity operator, lacO2, in the following way: During growth on glucose the binding site of LacR to the lacO1 represses transcription of the lac promoter while activates transcription of lacR. With increasing concentrations of LacR, the lac operon and lacR gene expression are repressed. This is due to the lower affinity of lacO2 for LacR than lacO1. Binding of the inducer to LacR, results in dissociation of the complex of LacR-operator complex and expression of the operon [1, 3, 34]. The inducer is tagatose-6-phosphate generated during growth on lactose.

Although the lactose specific components of the PTS and the enzymes of the tagatose-6-phosphate pathway are plasmid-located in most *L. lactis* strains, there are cases of strains in which the genes were found to be chromosomally located [35, 36].

PTS and growth of glucose

The metabolism of lactose, glucose and galactose is of special importance to the dairy industry and all industries involved in production of microbial metabolites. In *Bacillus subtilis*, the glucose-specific PTS, comprising EI and HPr and the EII^{Glc} complex, plays an important role in transport and phosphorylation of glucose [37]. In lactic acid bacteria and sugar-fermenting streptococci, transport and phosphorylation of glucose is carried out mainly by the mannose PTS, phosphoenolopyruvate:mannose phosphotransferase system, (EI and HPr and the EII^{Man} complex) [38]. Various PTSs have been identified for a number of LAB [38], e.g. *Lactobacillus casei*, *Lb. sakei*, *Lb. curvatus*, and several species of oral streptococci, as well as for *L. lactis* [8, 12, 39, 40].

Kinetic analysis of the PTS-mediated transport of glucose in *S. lactis* ML3 has been carried out by Thompson [8]. The initial rates of uptake of glucose by intact cells displayed high-affinity Michaelis-Menten saturation characteristics. Transformation of the initial rate data according to the method of Hofstee, yielded the kinetic parameters V_{max}= 478 μmol/g (dry weight) of cells per min and K_m= 15.5 μM. Papagianni et al. [40] worked with the strain *L. lactis* spp. *lactis* LM0230 in studies of the relationship between the glycolysis and the regulation of glucose transport in aerated cultures. Kinetic analysis of the PTS-mediated transport system of glucose, performed according o Thompson [8], produced again initial rates of glucose uptake with high-affinity Michaelis-Menten characteristics. However, transformation of the data according to Eadie-Hofstee yielded the following kinetic parameters: V_{max}= 107 mmol $min^{-1}g^{-1}$ and K_m= 2mM, which are significantly different from the reported by Thompson [8] for a different strain. In the same study [40], the presence of a low-affinity carrier was reported for the first time. That appeared also to be involved in glucose transport at higher glucose concentrations (27.5-55 mM) and was found to be characterized by the following parameters: V_{max}= 278 mmol $min^{-1}g^{-1}$ and K_m= 14mM. Solving the Michaelis-Menten equation $V = \frac{V_{max} \cdot S}{K_m + S}$ for the estimated V_{max} and K_m values at various glucose concentrations within a wide range (13.75-555 mM), the quoted units for V were converted to specific uptake rates and plotted along with the experimentally derived values for specific uptake rates. The methodology revealed that the experimentally derived values for specific uptake rates were higher than the calculated with the mediated high-affinity transport model. At glucose concentrations between 27.5 and 55 mM, glucose was transported by a low-affinity carrier, while at even higher glucose levels, accumulation of unphosphorylated glucose inside the cells was explained as a result of uncontrolled glucose entry by unfacilitated (simple) diffusion.

The EII^{Man} complex plays a major role in glucose transport and phosphorylation in LAB and can be assumed that the activity of this PTS would affect catabolite repression (CR) [38]: Mutations rendering the EII^{Man} complex inactive, resulted in the loss of the preferential use of glucose over several carbon sources, such as lactose or ribose in *Lb. casei* and other LAB [38, 41, 42]. In several cases, a regulatory role in CR has been suggested for the EII^{Man} complex but in overall the mechanisms by which the complex is implicated in regulatory functions are not satisfactory defined [38].

Glucose is transported inside the cell mainly by the mannose–PTS and once internalized it is phosphorylated by EIIA to glucose-6-phosphate to enter the glycolytic pathway. The mannose-PTS system apart form glucose, transports also mannose, fructose, glucosamine, and 2-deoxy-D-glucose. For some strains however, another PTS system has been described, the glucose-PTS that exhibits specificity to glucose and α-methyl-glucoside [10].

PTSs for Other Sugars and Other than PEP:PTS Sugar Transport Systems in Lactococci

Fructose and sucrose are important sugars in the food industry. Fructose can be transported either by the mannose-PTS, yielding fructose-6-phospate, or by a specific fructose-PTS, and the resulting fructose-1-phosphate enters glycolysis as FBP after phosphorylation [43]. Sucrose uptake in some *L. lactis* strains is mediated by a sucrose-PTS [9], and the resulting sucrose-6-phosphate is hydrolyzed (by sucrose-6-phosphate hydrolase) to glucose-6-phosphate and fructose. A specific trehalose-PTS system has also been discovered and described by Andersson et al. [44]. β-phosphoglucomutase is involved in the metabolism of trehalose, which enters the cell and it is converted to glucose-6-phosphate and β-glucose-1-phosphate via trehalose-6-phosphate phosphorylase.

Sugar transport via secondary systems (permeases) is coupled to ion translocation, and is followed by a kinase-mediated phosphorylation step [15]. The secondary transport system for lactose was the first ion-linked transport system reported for *L. lactis* [45]. Since then, a number of secondary systems for sugar transport in *L. lactis* have been described, belonging to the galactoside-pentose-hexuronide group of transport systems [46]. An ATP-dependent permease system has been described for maltose [47, 48, 49]. Through the action of a P_i-dependent maltose phosphorylase, maltose is converted to glucose and β-glucose-1-

phosphate. Anomerization of the latter into α-glucose-1-phosphate and entry into glycolysis is via the action of a specific β -phosphoglucomutase [50].

The genes that encode for most of the above-mentioned proteins have been identified in the genome sequence of *L. lactis* IL1403 [13]. Still remain unknown however, the genes coding for glucose permeases and a galactose-specific PTS [51].

A large number of transporter proteins in *L. lactis*, ATP-dependent, ion-channel, PTS-specific transporters, secondary and various unclassified transporters are known today and published in databases, e.g. the one maintained by J. Craig Venter Institute, USA, (at www.membranetransport.org). The following PTS transporters are known today for *L. lactis* IL1403: General PTS: L120335 (ptsH), subtype HPr and L120628 (ptsI), subtype Enzyme I; Sugar-specific PTS: L145238 (yleD, Enzyme IIBC) and L1762179 (ptnAB, Enzyme IIAB) for sucrose, L147466 (ptnD, Enzyme IID) and L146623 (ptnC, Enzyme IIC) for mannose, L177520 (celB, EnzymeIIC), L19292 (ptcA, EnzymeIIC), L20847 (ptcC, EnzymeIIC) and L31294 (yidB, EnzymeIIC) for cellobiose, L32907 (mtlF, EnzymeIIA) for mannitol, L185031 (fruA, EnzymeIIABC) and L18872 (ptcB, EnzymeIIB) for fructose, L146642 (yleE, EnzymeIIABC), L37906 (yedF, EnzymeIIABC), and L90678 (ptbA, EnzymeIIABC) for β-glucosides. Examples of known ATP-dependent transporters are the following: L129753, which contains both a membrane domain and a binding protein domain as one polypeptide, L128777 and L27865 (multiple sugar transporters). Transporters of the Glycoside-Pentoside-Hexuronid (GPH):Cation Symporter Family are the L0023 (uxuT), a sodium ion: galactoside transporter, the L0233 (xynT), a proton-sodium ion:xylose transporter and the L113994 (ypbD), a proton-sodiumion:sugar transporter. Also, a specific transporter involved in the uptake of glucose is the L140621 (yxfA) which belongs to the Drug/Metabolite Transporter (DMT) Superfamily of transporters [52].

Figure 1, gives a summary of the main pathways involved in transport and initial metabolism of mono- and disaccharides in *L. lactis*.

Glycolysis

The main purpose of sugar metabolism in *L. lactis*, a facultative anaerobe and homofermentative lactic acid bacterium, is to produce ATP for biosynthesis. The free energy metabolism of *L. lactis* is rather simple. During fermentation, more than 95% of the substrate ends up in fermentation products. The main metabolic product is lactate. The fermentation pattern shows that the role of glycolysis is to supply ATP for growth and maintenance (figure 2). Oxidative phosphorylation does not normally occur in *L. lactis* and ATP is generated by glycolysis. Glucose is converted to pyruvate through glycolysis, with production of ATP by substrate level phosphorylation and reducing equivalents (NADH) at the level of glycaraldehyde-3-phosphate dehydrogenase (figure 2). Reduction of pyruvate to lactate via the enzyme of lactate dehydrogenase (LDH) maintains the redox balance by generating NAD^+. Accumulation of elevated levels of FBP (fructose bi-phosphate) is a major characteristic of glucose metabolism in *L. lactis* [53, 54].

It can be considered, if the small amount of generated NADH in anabolism is neglected, that catabolism is constrained by a balance between NADH-producing and NADH-consuming reactions. The result, under anaerobic conditions, is the conversion of glucose into lactate (LDH) or into the mixed acid products formate, ethanol, and acetate at a molar ratio of 1:1:1 via PFL (pyruvate formate lyase) depending on whether the specific sugar uptake is high or low [55, 56, 57]. Therefore, mixed acid products accumulate in only low quantities during homolactic fermentation but have been shown to account for the majority of carbon flux under conditions in which the rates of glycolysis of sugars are very low. Under aerobic conditions, the tight coupling of catabolic carbon fluxes that is needed to satisfy the redox balance is alleviated and NAD^+ can be regenerated by the activity of NADH oxidases (NOX).

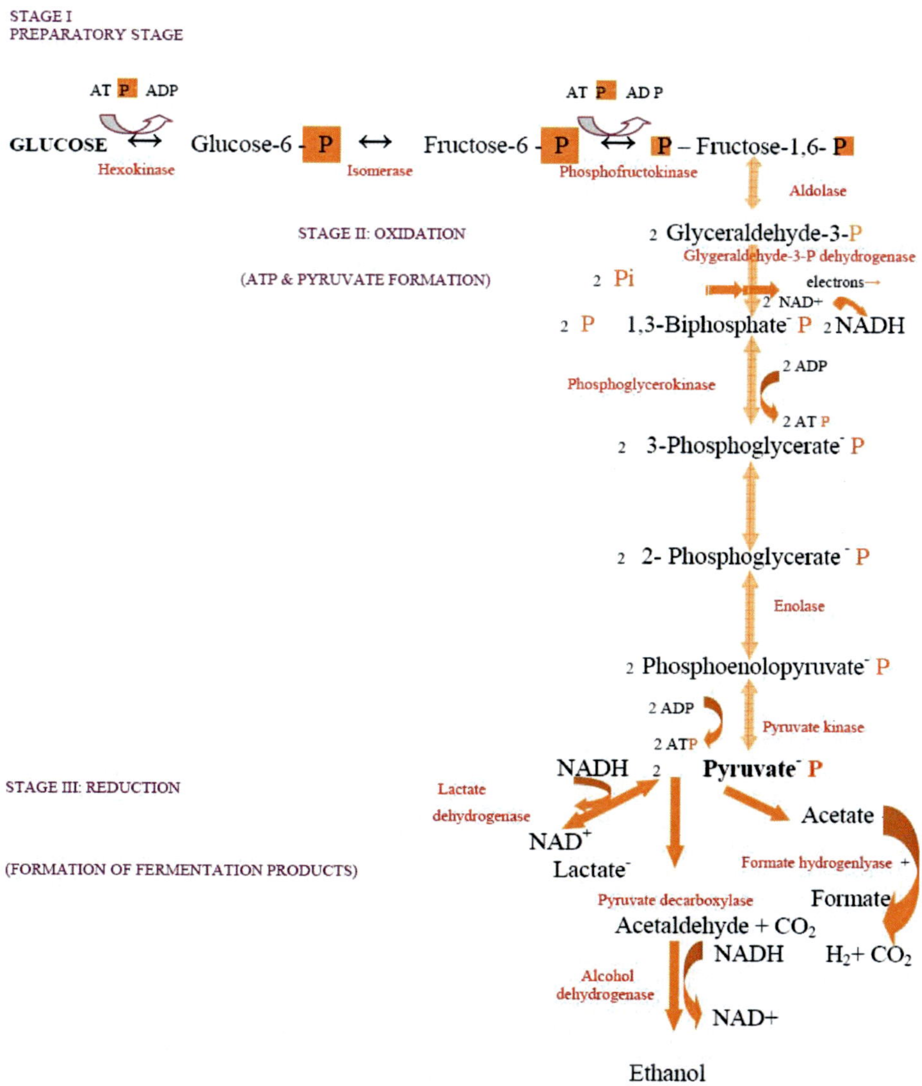

Figure 2. Glycolysis (Embden-Meyerhof) pathway, the sequence of enzymatic reactions in the conversion of glucose to pyruvate and finally, to fermentation products. In red letters, are the enzymes involved. Highlighted, are the components exchanged between oxidation or reduction reactions. The number of the produced molecules is given, highlighted in green.

Sugar Substrate

Luesink et al. [19, 26] showed that growth on glucose resulted in higher activities of the key glycolytic enzymes phosphofructokinase (PFK), pyruvate kinase (PK), and also L-lactate dehydrogenase (LDH), the genes of which form the tricistronic *las* operon. Although sugar metabolism is the most important issue in *L. lactis* physiology studies, growth on glucose as the sole carbon source is the case for a small only number of studies [9, 58, 59, 60, 61], the majority carried out mainly with lactose. Even et al. [58], using a novel DNA macroarray technology, showed that several genes of glycolysis were expressed to higher levels on glucose and that the genes of the mixed acid pathway were expressed to higher levels on galactose.

Even et al. [58] reported data on specific rates of growth, substrate consumption and product formation (lactate, acetate, formate, and ethanol) during growth of *L. lactis* IL1403 on two different synthetic media (MCD and MS10R) with glucose and galactose as carbon sources. Glucose supported higher specific growth rates, higher sugar consumption rates and lactate production rates in both media than galactose. Specific production rates for formate, acetate and ethanol were comparable in the two substrates. Glycolytic enzymes PFK, PK, and the LDH specific activities were higher in both media with glucose than with galactose.

Not only the type, but also the concentration of the sugar substrate influences the overall fermentation rates and productivities. Papagianni et al. [40], carried out batch and fed-batch experiments with *L. lactis* spp. *lactis* LM0230 in a stirred tank bioreactor, under microaerobic conditions and a range of glucose concentrations from 13.75 to 555 mM. The tool of glucostat fed-batch culture was employed, in which glucose was added at a rate suitable to maintain a stable concentration throughout the runs. In batch culture, the initial glucose concentration of 138 mM supported the highest specific rates for growth, glucose uptake and lactate production. In fed-batch culture, maximal rates obtained by maintaining a continuous 55 mM glucose concentration. The maximum values obtained in batch runs occurred when the sugar had fallen to values that according to glucostat data were too low to give significant lactate production. The derived data indicated that there must be two aspects to the effect of sugar. One is the level itself, the other arising from the dynamic situation with cells being exposed to a constantly changing glucose level in the bioreactor.

Maintenance of a continuous low glucose level, such as 13.75 mM in fed-batch culture in the same study [40], resulted in low specific glucose uptake rates and in a shift towards mixed acid metabolism. It is well-known for anaerobic culture that homolactic metabolism occurs during cultivation in substrates that support rapid growth, in which significant amounts of glucose remain in the medium, while mixed acids metabolism occurs when growth rates are rather low and in true carbon-limited chemostats [56, 62]. Working the glucostat fed-batch culture mode under microaerobic conditions, a situation was arranged in which significant amounts of glucose were always present in the fermentation broth with glucose being the substrate supporting the highest fermentation rates. Therefore, the shift from homolactic to mixed acid fermentation could be directly correlated to the glucose uptake rate and consequently, to the flux through glycolysis.

Specific activities of the PFK, PK, and LDH, were found to be strongly influenced by the level of glucose in glucostat fed-batch experiments [40]. The 55 mM glucose level supported the highest enzyme activities and this was mirrored on the intracellular metabolites pools. Low specific enzyme activities were obtained in the presence of high glucose levels, e.g. 277 mM, while the maximum glycolytic flux of 25.5 mmol g $CDW^{-1}h^{-1}$ was observed in the 55 mM glucostat. The observed 55% reduction in the glycolytic flux corresponded to a 56% reduction of PFK activity. The explanation for the negative influence of elevated glucose levels on the glycolytic flux is likely to lie in part in the depressed *pfk* gene activity.

Anaerobic *vs* Aerobic Growth

Most studies with *L. lactis* have been carried out under anaerobic conditions. In few cases only, the conditions were fully aerobic and to the best of our knowledge these were the works of Jensen et al. [63], Lopez de Felipe [64], Cogan et al. [65], and Van Neil et al. [66]. Intermediate oxygen concentrations have been applied in the works by Nordkvist et al. [61] and Jensen et al. [63], both of which were carried out with *L. lactis* spp. *cremoris*, and the work by Papagianni et al. [40] with *L. lactis* spp. *lactis*. These studies were carried out under microaerobic conditions, 5% dissolved oxygen tension (DOT) relative to saturation with air and with glucose as the sole carbon source. Comparisons at different aeration levels were made in the works of Nordkvist et al. [61] for *L. lactis* spp. *cremoris*, and Papagianni et al. [40] for *L. lactis* spp. *lactis*. In both cases, the maximum specific growth rate decreased with increasing aeration, while an optimum yield of lactate on glucose was obtained under microaerobic conditions compared to anaerobic, fully aerobic, and semiaerobic (50% DOT).

The ability of *L. lactis* to grow under aerobic conditions has been correlated with the presence of the flavoproteins NADH oxidase and NADH peroxidase, an H_2O-forming NADH oxidase, and a manganese-containing superoxide dismutase (SOD) [67, 68, 69, 70, 71]. Intensive research in the area has identified a key role for the $NADH/NAD^+$ ratio (or the internal redox state) in the regulation of sugar metabolism [56, 63, 72, 73, 74]. Neves et al. [74], by using a ^{13}C NMR *in vivo* showed that the glycolytic flux decreased in the presence of saturating levels of oxygen, but it was not altered in response to changes in the NADH oxidase activity. In that study, three isogenic strains of *L. lactis* were used: the parent *L. lactis* MG1363, a NOX^- strain harboring a deletion of the gene encoding the H_2O-forming NADH oxidase, and a NOX^+ strain with the NADH oxidase activity enhanced by about 100-fold. The observation that the glycolytic flux was not enhanced in the last case of the NOX^+ strain indicated that the glycolytic flux was not primarily determined by the level of NADH in the cell. An explanation was given to the phenomenon of the negative effect exerted by oxygen on the glycolytic flux that this is likely to lie in part in the depressed activity of *pfk* gene.

Aeration has also been shown to strongly influence the cellular content of key enzymes. The negative effect of oxygen on the expression of *pfl* gene that encodes the enzyme of pyruvate formate lyase is well-known [75, 76]. The *pfl* gene has been shown to be very sensitive to oxygen [67, 76, 77]. Another gene, the expression of which is well-known to be affected by oxygen, is the *adh* gene that encodes for the alcohol dehydrogenase enzyme [78]. The levels of the key glycolytic enzymes PFK, PK, and the LDH were found to be reduced with increasing aeration [40, 79]. In contrast, the *in vitro* specific activities of α-acetolactate synthase (ALS) and the pyruvate dehydrogenase (PDH) complex have been reported to increase with aeration [63, 65].

Regulation of Glycolysis

The regulation of glycolysis and the shift between the various fermentation modes in *L. lactis* have been subjects of extensive research [39, 53, 58, 59, 63]. The mapping of regulatory mechanisms, however, does not necessarily lead to an understanding of which enzymes have control on the flux [80]. Prior knowledge of the metabolic pathways and more recently, of the genome sequence of *L. lactis* has led to successful application of modulation of gene expression and Metabolic Control Analysis (MCA) [80], as well as *in vivo* NMR [81] and various cloning techniques in investigations on the regulation of glycolysis in this organism.

In MCA, flux control by an enzymatic reaction step can be determined by changing the activity of the enzyme away from the normal and determining the effect on the metabolic flux. *Modulation* or *tuning* of gene expression is advantageous in order to perform MCA and various genetic tools are available today for *L. lactis*. With respect to nuclear magnetic resonance spectroscopy (NMR), the development of high-field superconducting magnets together with the emergence of the Fourier transform NMR method, revolutionized the scope of the technique and allowed researchers to apply and benefit from the capabilities of NMR through carrying out measurements directly on living systems. ^{13}C NMR is the technique of choice in most cases because of its large chemical shift range. The major drawback of NMR however, is its intrinsic low sensitivity, which limits *in vivo* observations to metabolites present in mM concentrations (relatively high). The majority of NMR experiments are carried out with thick suspensions of non-growing cells.

Key Enzymes and Pools of Metabolites – Products of the Respective Enzymatic Reactions

The Las Operon Enzymes

The *las* operon harbours the three genes *pfk*, *pyk*, and *ldh* coding for phosphofructokinase (PFK), pyruvate kinase (PK), and lactate dehydrogenase (LDH), respectively [82]. The *las* operon genes and their enzymes have been the focus of a large amount of research on the regulation and control of glycolysis.

In an attempt to change the expression of the *las* operon in *L. lactis*, Andersen et al. [83] used the synthetic promoters constructed by Jensen and Hammer [84]. Two constitutive promoters, each flanked by the upstream region of the *las* operon and the truncated *pfk* region were cloned on an *E. coli* vector and the plasmids were transformed in *L. lactis* spp. *cremoris* MG1363, resulting in construction of two strains in which two synthetic constitutive promoters with different strengths had replaced the native *las* promoter. The *las* mutants were found to have uncoordinated expression of the *pfk*, *pyk*, and *ldh* genes relative to the wild-type strain. While the constructed strains had an almost two-fold decrease in PFK activity,

PK and LDH activities remained closer to the wild-type level. The lower PFK activity resulted in reduction of the growth rate and a proportional reduction of the glycolytic flux. The later phenomenon is a strong indication of the critical role of the PFK in controlling the glycolytic flux. However, conclusions about flux control could not be drawn directly from these experiments – modulation of the expression of PFK activity instead was required. The elevated pools of the hexose phosphates were indicative of the PFK control over the concentration of the upstream metabolites.

Determination of the specific activities of the key glycolytic enzymes PFK, PK and LDH in the work of Papagianni et al. [40], showed that expression of the *las* operon genes in microaerobic glucostat fed-batch cultures was influenced by the glucose level. The 55 mM glucose level supported the highest enzyme activities (within the tested range of glucose levels of 13.75 to 555 mM) and this was reflected on the intracellular metabolites pools. The cource of FBP (fructose-1,6 bisphosphate, the product of PFK reaction) concentration over increasing levels of glucose and the intracellular accumulation of unphosphorylated glucose were suggested to be indicative of repressed PFK activity. As it has been mentioned earlier, a strong influence of PFK on the glycolytic flux was identified in the works by Andersen et al. [83] and Neves et al. [74], in studies with the level of oxygen. A different approach in the work of Papagianni et al. [40], through the glucose level, demonstrated and validated the regulatory role of PFK on glycolytic flux in *L. lactis*.

Accumulation of FBP to high levels (around 50 mM) is a major characteristic of glucose metabolism in *L. lactis* [81]. The finding that FBP is an allosteric regulator (activator) of PK and LDH suggested that it plays an important role in regulation of *L. lactis* metabolism [53, 83, 84]. High levels of FBP activate PK and LDH and direct the flux towards lactate production, while low levels of FBP lead to LDH inactivation and inhibition relief of pyruvate-formate lyase by triose phosphates, resulting to a shift to mixed acid fermentation. Garrigues et al. [56] have questioned such a direct effect, since the detected intracellular concentrations of FBP are in general sufficient to ensure full activation of LDH. Also, more recently in the work of Papagianni et al. [40] it has been shown that the concentrations of FBP pools and the $NADH/NAD^+$ ratio in the glucostat runs of 13.75 and 138 mM glucose were almost identical, while neither the specific glucose uptake rates nor the fermentation pattern (mixed acid, homolactic, respectively), were similar. Moreover, much reduced FBP pools at even higher glucose levels in the glucostat suggest that FBP cannot be regarded as a direct regulator of product formation, while they provide an indication of inhibition of PFK activity at such high glucose levels. While the FBP pool level cannot be directly connected to the glycolytic flux and the fermentation pattern, the explanation of the phenomena rather lies in the ATP demand of the cells and the glucose transport capacity of the microorganism. In their review, Neves and co-workers [81] noted that it is possible that the role of FBP as regulator was overestimated because of its relatively elevated concentrations that can be easily measured compared to those of other intracellular metabolites.

FBP, however, was shown recently to be a major signaling molecule for carbon catabolite protein A (CcpA)-dependent catabolite repression and activation of genes in Gram-positive bacteria [81]. Phosphorylation of HPr at Ser-46 is mediated by the bifunctional enzyme HPr kinase/phosphorylase (HPrK/P); the kinase activity of HPr is allosterically activated by FBP and inhibited by Pi, which serves as a substrate for the phosphatase reaction

[85, 86]. Therefore, FBP and Pi, the main regulators of sugar metabolism in *L. lactis*, in part due to their dual but antagonistic modulation of PK activity, were shown to be critical factors in a global control mechanism. FBP may provide a link between glycolytic activity and carbon catabolite repression in Gram-positive bacteria.

CcpA was also found to be a transcriptional activator of the *las* operon, modulating glycolytic activity by controlling the key enzymes PFK, PK and LDH. Enhancement of the binding of CcpA to *cre* sites in response to FBP, though suggested, has not yet been proven [19, 26, 81].

The level of FBP pool is high in energized cells, but the force that drives the accumulation of this metabolite still remains the subject of discussion [81]. Garrigues et al. [56] suggested that inhibition or activation exerted by the ratio of $NADH/NAD^+$ on glyceraldehyde phosphate dehydrogenase (GAPDH) or LDH is the main issue regulating glycolysis. In this work, the shift from homolactic to mixed acid fermentation in *L. lactis* has been directly correlated to the glycolytic flux, estimated from the specific rates of sugar (glucose, galactose and lactose) consumption. Under anaerobic conditions, the predominant role of $NADH/NAD^+$ ratio in controlling the shift was shown, as well as the relationship between GAPDH activity and the $NADH/NAD^+$ ratio. However, under conditions supporting less rapid growth, with a diminished flux through glycolysis and a lower $NADH/NAD^+$ ratio, such as growth on galactose or lactose, the major pathway bottleneck was identified at the level of sugar ransport rather than GAPDH. The influence of GAPDH on glycolysis has been discussed as either strictly controlling [87] or having such a role only under conditions of high glycolytic flux [56]. Quite different regulatory aspects of glucose metabolism in the presence of oxygen have been reported by Neves et al. [79]. These investigators showed that the glycolytic flux was not primarily determined by the level of NADH in the cell. The main point in their work was the observation that the decrease in the level of PFK activity by 40% was proportional to the decrease in the glycolytic flux. A negative effect of oxygen on the flux through glycolysis was identified and explained by depressed PFK activity. The same group, working with another srain of *L. lactis* (MG5627) [51], observed a stimulation of glucose assimilation under semiaerobic conditions, a fact characterized by them as "apparent discrepancy" which showed that the level of oxygen notably affected the cell metabolic machinery through different effects on gene expression.

The approach of modulating gene expression via synthetic promoters has been used, apart from PFK, to study the importance of LDH for metabolic fluxes in *L. lactis* MG1363. A full version of the *ldh* gene was cloned behind a set of constitutive promoters in a plasmid vector that allowed for site specific integration in a phage attachment site on the chromosome [88]. The vector was introduced into the strain and into a version of it with disrupted *ldh* gene, resulting in a series of mutant strains with modulated LDH activities, ranging from 1 to 133% of the wild-type level [89]. No effect was observed on the glycolytic flux and the growth rate through changing the LDH activity from 59 to 133% of wild-type level. Determination of the flux control coefficients showed that LDH had no control on the growth rate, glycolytic flux and lactate production but had a strong negative control on the flux to formate.

The Impact of Oxygen

L. lactis is mostly studied under anaerobic conditions and it is regarded as a facultative anaerobe. Genome analysis of *L. lactis* spp. *lactis* IL1403 [13], however, indicated the presence of almost all functions needed for aerobic respiration in this microorganism. It possesses *men* and *cyt*ABCD operons, encoding the proteins required for the synthesis of menaquinone and cytochrome *d* and also three genes involved in the late steps of heme synthesis but not the genes required for the early steps. It was observed that during growth under fully aerobic conditions, addition of heme leads to diauxic growth, improvement of biomass yield and long-term survival; fermentation occurs first, and it is followed by respiration that occur with the depletion of glucose. Increased biomass yields under aerobic conditions without addition of hoxegenous heme, have been obtained with two different strains, *L. lactis* LM0230 and *L. lactis* ATCC 11454 in the works of Papagianni et al. [40, 90].

The effect of oxygen on the distribution of end products in *L. lactis* fermentation has long been discussed, but its impact on the glycolytic metabolite pools was investigated only during the last decade. Neves et al. [79], carried out in vivo ^{13}C NMR analysis of non-growing cell suspensions to obtain a more reliable picture of the oxygen induced changes in glycolytic metabolite pools. The maximum level of FBP and the rate of its consumption, and the 3-PGA and PEP pools were increased in the presence of oxygen. Under an oxygen atmosphere, the NADH oxidase provides an additional path for NADH oxidation and the lower FBP accumulation is due to the increase of the flux through GAPDH caused by the lower NADH concentration levels. The same reasoning was applied to explain the faster FBP consumption. GAPDH could sustain a higher flux, since the enzyme was less inhibited by the lower NADH levels. *In vivo* NMR at the onset of glucose exhaustion revealed no NADH accumulation in the presence of oxygen. At that metabolic stage, accumulation of 3-PGA and PEP is driven by PK inhibition. Thus, under aerobic conditions, NADH consumption by NADH oxidase obviates the need to regenerate NAD^+ downstream of pyruvate and to overcome the PK bottleneck. This way, 3-PGA and PEP that derive from the metabolism of residual FBP accumulate at elevated levels.

CcpA was found to be involved in the regulation of the shift from fermentation to respiration, by controlling both expression of *noxE*-encoding NADH oxidase and heme uptake [91]. CcpA-mediated repression of *noxE* has more metabolic consequences, since it refers to the redox status ($NADH/NAD^+$) as an important regulator of carbon metabolism in the presence of oxygen. Therefore, involvement of CcpA suggests a strong role of FBP in the overall regulation process.

ATP-Consuming Processes

The control of the flux through a pathway can also reside in processes outside the pathway itself, for example in processes that consume its products [80]. Using this approach, the demand for ATP was tested by modulating the activity of ATPase [80]. Increasing the expression of ATPases led to uncoupled biomass production from glycolysis and a lower

ATP/ADP ratio (strain MG1363). The glycolytic flux was determined in growing and non-growing cells and interestingly, it was found that it was not increased in the first case while it was 3-fold stimulated in the second case. The lower glycolytic flux with non-growing cells is due to the fewer ATP-consuming reactions. Under growing conditions, the glycolytic flux reached maximal levels and therefore, expression of ATPase resulted in increased flux.

According to Koebmann and co-workers [80], the demand for ATP exerts some control when the glycolytic flux is significantly lower than the maximal capacity. It becomes obvious that the glycolytic flux is distributed over many steps and in combination with ATP-consuming reactions. The process of sugar transport, although neglected by most investigators when the regulation of glycolysis is studied, deserves a critical role in the phenomena. Our research [40], using the tool of glucostat fed-batch culture, revealed that under microaerobic conditions (5% DOT) and during growth on glucose, the control of the glycolytic flux resides to a large extend in processes outside the pathway, like the ATP consuming reactions and glucose transport. Depending on culture conditions, e.g. dissolved oxygen concentration and glucose concentration levels, the overall flux in *L. lactis* seems to be regulated by the ATP demand through the allosteric properties of key enzymes, with PFK having a significant influence on the control. Following extensive metabolic analysis in growing cells of *L. lactis*, we proposed a regulation mechanism governed by the energy state of the cell, as this is expressed by the cellular quantities of ADP and ATP, through which *L. lactis* can handle the glycolytic flux under microaerobic conditions. ADP and ATP play central roles in the in metabolism and influence several steps of the glycolytic pathway since they are substrates and products of kinases and inhibitors of dehydrogenases. ATP acts as a free-energy donor to drive transport and bisosynthesis and it is continuously regenerated from ADP by substrate level phosphorylation. ATP is invested in the upper part of the pathway to generate a surplus in the lower part. Additionally, both ATP and ADP serve as precursors in DNA and RNA synthesis, which have been shown to constitute an about 3 and 8% of *L. lactis* dry biomass, respectively [92]. It has also been shown that intracellular concentrations of ADP and ATP in growing *L. lactis* cells are tightly controlled (homeostatic control) at levels optimal for the cellular reactions [92]. In our study, under low glucose concentration conditions provided in glucostat cultures, the glycolytic flux could not meet the anabolic demand of the cells. There was glucose limitation and consequently energy limitation and the glucose transport capacity of the microorganism was not met, resulting in mixed acids formation. The FBP pool, through LDH and PYK control, does not directly influence product formation since low FBP concentrations were characteristic of both low (13.75 mM) and high (138 mM) glucose concentration levels in glucostat cultures. Therefore, under such conditions, the ATP demand and the glucose transport capacity of the cells are main regulators of the flux. Providing constant elevated glucose levels in the glucostat (e.g. above 55 mM), conditions in which glucose transport carriers are saturated, led to excess energy and formation of large intracellular pools of ADP and ATP, which the organism can handle through the allosteric properties of its enzymes. Excess ATP in this case, inhibits PFK activity slowing the glycolytic flux down. It can be suggested here that the extent to which ATP demand controls the glycolytic flux depends on how much excess capacity of glycolysis is present at cells.

Concluding Remarks

Despite the large amount of information on *L. lactis* metabolism and the glycolytic pathway, it is not clear yet what controls the glycolytic flux. Many environmental parameters exert strong influence on gene expression and isolation of the phenomena cannot contribute to an overall understanding of the physiology of the microorganism. The available quantitative information from genomics and metabolomics research needs to be integrated into a dynamic model for at least one industrially important strain.

References

[1] Vaughan, EE; Kleerebezem M; de Vos WM. Genetics of the metabolism of lactose and other sugars. In: Wood, BJB; Warner, P (Eds.), Genetics of lactic acid bacteria. New York, USA: Kluwer Academic / Plenum Publishers; 2003; pp. 95-119

[2] McKay, LL. Regulation of lactose metabolism in dairy streptococci. In: Davies, R. (Ed.), Developments in food microbiology. London: Elsevier Applied Science Publishers; 1982; Vol. 1, pp. 153-182

[3] van Rooijen, RJ; de Vos, WM. Purification of the *Lactococcus lactis* LacR repressor gene and characterization of its DNA binding. In: van Rooijen, RJ (Ed.), Characterization of the *Lactococcus lactis* lactose genes and regulation of their expression. Ph.D. Thesis, Wageningen Agricultural University, The Netherlands; 1993; pp. 101-118

[4] de Vos, WM; Vaughan, EE. Genetics of lactose utilization in lactic acid bacteria. *FEMS Microbiol. Rev.* 1994, 15, 217-237

[5] Grossiord, B. Métabolisme du galactose par la voie de Leloir: l'óperon *gal* de *Lactococcus lactis*. Ph.D. Thesis. École Nationale Supérieure Agronomique de Montpellier, France, 1998

[6] McKay, LL; Walter, LA; Sandine, WE; Elliker, PR. Involvement of phosphoenolopyruvate in lactose utilization by group N streptococci. *J. Bacteriol.* 1969, 99, 603-610

[7] McKay, LL; Miller III, A; Sandine, WE; Elliker, PR. Mechanisms of lactose utilization by lactic acid streptococci: enzymatic and genetic analysis. *J. Bacteriol.* 1970, 102, 804-809

[8] Thompson, J. In vivo regulation of glycolysis and characterization of sugar:transferase systems in *Streptococcus lactis*. *J. Bacteriol.* 1978, 136, 465-476

[9] Thompson, J; Chassy, BM. Uptake and metabolism of sucrose by *Streptococcus lactis*. *J. Bacteriol.* 1981, 147, 543-551

[10] Thompson, J; Saier, MH. Regulation of methyl-β-D-thiogalactopyranoside-6-phosphate accumulation in *Streptococcus lactis* by exclusion and expulsion mechanisms. *J. Bacteriol.* 1981, 146, 885-894

[11] Thompson, J; Chassy, BM. Regulation of glycolysis and sugar phosphotransferase activities in *Streptococcus lactis*: Growth in the presence of 2-deoxy-D-glucose. *J. Bacteriol.* 1983, 154, 819-830

[12] Thompson, J; Chassy, BM. Intracellular phosphorylation of glucose analogs via the phosphoenolopyruvate: mannose-phosphotransferase system in *Streptococcus lactis. J. Bacteriol.* 1985, 162, 224-234

[13] Bolotin, A; Wincker, P; Mauger, S; Jaillon, O; Malarme, K; Weissenbach, J; Ehrlich, SD; Sorokin, A. The complete genome sequence of the lactic acid bacterium *Lactococcus lactis* ssp. *lactis* IL1403. *Genome Res*. 2001, 11, 731-753

[14] Postma, PW; Lengeler, JW; Jacobson, GR. Phosphoenolopyruvate:carbohydrate phosphotransferase systems of bacteria. *FEMS Microbiol. Rev.* 1993, 57, 543-594

[15] Poolman, B. Energy transduction in lactic acid bacteria. *FEMS Microbiol. Rev.* 1993, 12, 125-147

[16] Fath, MJ; Kolter, R. ABC transporters: bacterial exporters. *FEMS Microbiol. Rev.* 1993, 57, 995-1017

[17] Hengstenberg, W; Reiche, B; Eiserman, R; Fischer, R; Kessler, U; Tarrach, A; de Vos, WM; Kalbitzer, HR; Glaser, S. Structure and function of proteins involved in sugar transport by the PTS of Gram-positive bacteria. *FEMS Microbiol. Rev.* 1989, 63, 35-42

[18] Saier, MH; Reizer, J. Proposed uniform nomenclature for the proteins and protein domains of the bacterial phosphoenolopyruvate: Sugar phosphotransferase system. *J. Bacteriol.* 1992, 174, 1433-1438

[19] Luesink, EJ; Beumer, CMA, Kuipers, OP; de Vos, WM. Molecular characterization of *Lactococcus lactis* ptsHI operon and analysis of the regulatory role of HPr. *J. Bacteriol.* 1999, 181, 764-771

[20] Leboeuf, C; Auffray, Y; Hartke, A. Cloning, sequencing and characteriztion of the *ccpA* gene from *Enterococcus faecalis*. *Int. J. Food Microbiol.* 2000, 55, 109-113

[21] Leboeuf, C; Leblanc, L; Auffray, Y; Hartke, A. Characterization of the ccpA gene of *Enterococcus faecalis*: identification of starvation-inducible proteins regulated by *ccpA*. *J. Bacteriol.* 2000, 182, 5799-5806

[22] Gunnewijk, MG; Poolman, B. Phosphorylation state of HPr determines the level of expression and the extent of phosphorylation of the lactose transport protein of *Streptococcus thermophilus*. *J. Biol. Chem.* 2000, 275, 34073-34079

[23] Titgemeyer, F; Hillen, W. Global control of sugar metabolism: a Gram-positive solution. *Antonie van Leeuwenhoek* 2002, 82, 59-71

[24] Ye, JJ; Reizer, J.; Cui, X; Saier, MH. ATP-dependent phosphorylation of serine-46 in the phosphocarrier protein HPr regulates lactose/H+ symport in *Lactobacillus brevis*. *Proc. Natl. Acad. Sci.* U.S.A. 1994, 91(8), 3102-3106

[25] Ye, JJ; Reizer, J.; Cui, X; Saier, MH. Inhibition of the phosphoenolopyruvate:lactose phosphotransferase system and activation of a cytoplasmic sugar-phosphate phosphatase in *Lactococcus lactis* by ATP-dpendent metabolite-activated phosphorylation of serine 46 in the phosphocarrier protein HPr. *J. Biol. Chem.* 1994, 269(16), 11837-11844

[26] Luesink, EJ; van Herpen, RE; Grossiord, BP; Kuipers, OP; de Vos, WM. Transcriptional activation of the glycolytic *las* operon and catabolite repression of the *gal* operon in *Lactococcus lactis* are mediated by the catabolite control protein CcpA. *Mol. Microbiol.* 1998, 30, 789-798

[27] Gasson, MJ. Plasmid complements of *Streptococcus lactis* NCDO712 and other lactic streptococci after protoplast-induced curing. *J. Bacteriol.* 1983, 154, 1-9

[28] Maeda, S; Gasson, MJ. Cloning, expression and location of the *Streptococcus lactis* gene for phospho-beta-D-galactosidase. *J. Gen. Microbiol.* 1986, 132, 331-340

[29] de Vos, WM; Gasson, MJ. Structure and expression of the *Lactococcus lactis* gene for phospho-beta-galactosidase (*lac*G) in *Escherichia coli* and *L. lactis*. *J. Gen. Microbiol.* 1989, 135, 1833-1846

[30] de Vos, WM; Boerrigter, I; van Rooijen, RJ; Reiche, B; Hengstenberg, W. Characterization of the lactose-specific enzymes of the phosphotransferase system in *Lactococcus lactis*. *J. Biol. Chem.* 1990, 265, 22554-22560

[31] van Rooijen, RJ; de Vos, WM. Molecular cloning, transcriptional analysis and nucleotide sequence of *lacR*, a gene encoding the repressor of the lactose phosphotransferase system of *Lactococcus lactis*. *J. Biol. Chem.* 1990, 265, 18449-18503

[32] van Rooijen, RJ; van Schalkwijk, S; de Vos, WM. Molecular cloning, characterization, and nucleotide sequence of the tagatose-6-phosphate pathway gene cluster of the lactose operon of *Lactococcus lactis*. *J. Biol. Chem.* 1991, 266, 7176-7181

[33] van Rooijen, RJ; Gasson, MJ; de Vos, WM. Characterization of the promoter of the *Lactococcus lactis* lactose operon: Contribution of flanking sequences and LacR repressor to its activity. *J. Bacteriol.* 1992, 174, 2273-2280

[34] van Rooijen, RJ; de Vos, WM. Nucleotide sequence of an iso-ISS*1* element flanking the 3 end of the *Lactococcus lactis* operon. In: van Rooijen, RJ (Ed.), Characterization of the *Lactococcus lactis* lactose genes and regulation of their expression. Ph.D. Thesis, Wageningen Agricultural University, The Netherlands; 1993; pp. 59-61

[35] Crow, VL; Davey, GP; Pearce, LE; Thomas, TD. Plasmid linkage of the D-tagatose 6-phosphate pathway in *Streptococcus lactis*: effect on lactose and galactose metabolism. *J. Bacteriol.* 1983, 153, 76-83

[36] de Vos, WM. Gene cloning and expression in lactic streptococci. *FEMS Microbiol. Rev.* 1987, 46, 281-295

[37] Gonzy-Treboul, G; de Waard, JH; Zagorec, M; Postma, PW. The glucose permease of the phosphotransferase system in *Bacillus subtilis*: evidence for JJGlc and III Glc domains. *Mol. Microbiol.* 1991, 5, 1241-1249

[38] Chailou, S; Postma, PW; Pouwels, PH. Contribution of the phosphoenolopyruvate:mannose phosphotransferase system to carbon catabolite repression in *Lactobacillus pentosus*. *Microbiol.* 2001, 147, 671-679

[39] Thompson, J. Sugar transport in lactic acid bacteria. In: Riezer, J; Peterkofski, A (Eds), Sugar transport and metabolism in Gram-positive bacteria. Chisester, UK: Ellis Horwood Limited; 1987; pp. 13-38

[40] Papagianni, M; Avramidis, N; Filiousis, G. Glycolysis and the regulation of glucose transport in *Lactococcus lactis* spp. *lactis* in batch and fed-batch culture. *Microbial Cell Factories* 2007, 6:16

[41] Veyrat, A. Monedero, V; Perez-Martinez, G. Glucose transport by the Phosphoenolopyruvate:mannose phosphotransferase system in *Lactobacillus casei*

ATCC 393 and its role in carbon catabolite repression. *Microbiol.* 1994, 140, 1141-1149

[42] Gosalbes, MJ; Monedero, V; Alpert, CA; Perez-Martinez, G. Establishing a model to study the regulation of the lactose operon in *Lactobacillus casei*. *FEMS Microbiol. Lett.* 1997, 148, 83-89

[43] Benthin, S; Nielsen, J; Villadsen, J. Two uptake systems for fructose in *Lactococcus lactis* subsp. *cremoris* FD1 produce glycolytic and gluconeogenic fructose phosphate and induce osclillations in growth and lactic acid formation. *Appl. Environ. Microbiol.* 1993, 59, 3206-3211

[44] Andersson, U; Levander, F; Radstrom, P. Trehalose-6-phosphate phosphorylase is part of a novel metabolic pathway for trehalose utilization in *Lactococcus lactis*. *J. Biol. Chem.* 2001, 276, 42707-42713

[45] Kashket, ER; Wilson, TH. Proton-coupled accumulation of galactosidase in *Streptococcus lactis* 7962. *Proc. Natl. Acad. Sci.* USA, 1973, 70, 2866-2869

[46] Poolman, B; Knol, J; van der Does, C; Henderson, PJ; Liang, WJ; Leblanc, G; Pourcher, T; Mus-Veteau, I. Cation and sugar selectivity determinants in a novel family of transport proteins. *Mol. Microbiol.* 1996, 19, 911-922

[47] Law, J; Buist, G; Haandrikman, A; Kok, J; Venema, G; Leenhouts, K. A system to generate chromosomal mutations to *Lactococcus lactis* which allows fast analysis of targeted genes. *J. Bacteriol.* 1995, 177, 7011-7018

[48] Levander, F; Andersson, U; Radstrom, P. Physiological role of beta-phosphoglucomutase in *Lactococcus lactis*. *Appl. Environ. Microbiol.* 2001, 67, 4546-4553

[49] Nilsson, U; Radstrom, P. Genetic localization and regulation of the maltose phosphorylase genes, *malP*, in *Lactococcus lactis*. *Microbiol.* 2001, 147, 1565-1573

[50] Qian, N; Stanley, GA; Hahn-Hagerdal, B; Radstrom, P. Purification and characterization of two phosphoglucomutases from *Lactococcus lactis* subsp. lactis and their regulation in maltose- and glucose-utilizing cells. *J. Bacteriol.* 1994, 176, 5304-5311

[51] Neves, AR; Pool, WA; Kok, J; Kuipers, OP; Santos, H. Overview on sugar metabolism and its control in *Lactococcus lactis* –The input from in vivo NMR. *FEMS Microbiol. Rev.* 2005, 29, 531-554

[52] Web site: www.membranetransport.org

[53] Thompson, J. Regulation of sugar transport and metabolism in lactic acid bacteria. *FEMS Microbiol. Rev.* 1987, 46, 221-231

[54] Ramos, A; Neves, AR; Santos, H. Metabolism of lactic acid bacteria studied by nuclear magnetic resonance. *Antonie van Leeuwenhoek* 2002, 81, 249-261

[55] Cocaign-Bousquet, M; Garrigues, C; Loubiere, P; Lindley, ND. Physiology of pyruvate metabolism in *Lactococcus lactis*. *Antonie van Leeuwenhoek* 1996, 70, 253-267

[56] Garrigues, C; Loubiere, P; Lindley, ND; Cocaign-Bousquet, M. Control of the shift from homolactic to mixed acid fermentation in *Lactococcus lactis*: predominant role of the NADH/NAD^+ ratio. *J. Bacteriol.* 1997, 179, 5282-5287

[57] Melchiorsen, CR; Jensen, NB; Christensen, B; Jokumsen, KV;Villadsen, J. Dynamics of pyruvate metabolism in *Lactococcus lactis*. *Biotechnol. Bioeng.* 2001, 74, 271-279

[58] Even, S; Lindley, ND; Cocaign-Bousquet, M. Molecular physiology of sugar catabolism in *Lactococcus lactis* IL1403. *J. Bacteriol.* 2001, 183, 3817-3824

[59] Garrigues, C; Coupil-Feuillerat, N; Cocaign-Bousquet, M. ; Renault, P ; Lindley, ND. Glucose metabolism and regulation of glucolysis in *Lactococcus lactis* strains with decreased lactate dehydrogenase activity. *Metab. Eng.* 2001, 3, 211-217

[60] Nielsen, J; Olsson, L. An expanded role for microbial physiology in metaboling engineering and functional genomics: moving towards systems biology. *FEMS Yeast Res.* 2002, 2, 175-181

[61] Nordkvist, M; Jensen, NBS; Villadsen, J. Glucose metabolism in *Lactococcus lactis* MG1363 under different aeration conditions: requirement of acetate to sustain growth under microaerobic conditions. *Appl. Environ. Microbiol.* 2002, 69, 3462-3468

[62] Thomas, TD; Ellwood, D; Longyear, M. Change from homo- to heterolactic fermenation by *Streptococcus lactis* resulting from glucose limitation in anaerobic chemostat cultures. *J. Bacteriol.* 1979, 138, 109-117

[63] Jensen, NBS; Melchiorsen, CR; Jokumsen, KV; Villadsen, J. Metabolic behavior of *Lactococcus lactis* MG1363 in microaerobic continuous cultivation at a low dilution rate. *Appl. Environ. Microbiol.* 2001, 67, 2677-2682

[64] Lopez de Felipe, F; Kleerebezem, M; de Vos, WM; Hugenholtz, J. Cofactor engineering: a novel approach to metabolic engineering in *Lactococcus lactis* by controlled expression of NADH oxidase. *J. Bacteriol.* 1998, 3804-3808

[65] Cogan, JF; Walsh, D; Condon, S. Impact of aeration on the metabolicend products formed from glucose and galactose by *Streptococcus lactis. J. Appl. Bacteriol.* 1989, 66, 77-84

[66] Van Neil, EWJ; Hahn-Hägerdal, B. Nutrient requirements of lactococci in defined growth media. *Appl. Microbiol. Biotechnol.* 1999, 52, 617-627

[67] Condon, S. Responses of lactic acid bacteria to oxygen. *FEMS Microbiol. Rev.* 1987, 46, 269-280

[68] Hansson, L; Häggstrom, MH. Effects of growth conditions on the activities of superoxide dismutase and NADH-oxidase/NADH-peroxidase in *Streptococcus lactis*. Curr. Microbiol. 1984, 10, 345-351

[69] Higuchi, M; Shimada, M; Yamamoto, Y; Hayashi, T; Koga, T; Kamio, Y. Identification of two distinct NADH oxidases corresponding to H_2O_2-forming oxidase and H_2O –forming oxidase induced in *Streptococcus mutan*s. *J. Gen. Microbiol.* 1993, 139, 2343-2352

[70] Lopez de Felipe, F; Starrenburg, MJC; Hugenholtz, J. The role of NADH-oxidation in acetoin and diacetyl production from glucose in *Lactococcus lactis* subsp. *lactis* MG1363. *FEMS Microbiol. Lett.* 1997, 156, 15-19

[71] Sanders, JW; Leenhouts, KJ; Haandrikman, AJ; Venema, G; Kok, J. Stress response in *Lactococcus lactis*: cloning, expression analysis, and mutation of the lactococcal superoxide dismutase gene. *J. Bacteriol.* 1995, 177, 5254-5260

[72] Hols, P; Ramos, J; Hugenholtz, J; Delcour, J; de Vos, WM; Santos, H; Kleerebezem, M. Acetate utilization in *Lactococcus lactis* deficient in lactate dehydrogenase: a rescue pathway for maintaining redox balance. *J. Bacteriol.* 1999, 181, 5521-5526

[73] Neves, AR; Ramos, A; Nunes, MC; Kleerebezem, M; Hugenholtz, J; de Vos, WM; Almeida, JS; Santos, S. *In vivo* nuclear magnetic resonance studies of glycolytic kinetics in *Lactococcus lactis*. *Biotechnol. Bioeng.* 1999, 64, 200-212

[74] Neves, AR; Ramos, A; Shearmen, C; Gasson, MJ; Almeida, JS; Santos, S. Metabolic characterization of *Lactococcus lactis* deficient in lactate dehydrogenase using in vivo ^{13}C-NMR. *Eur. J. Biochem.* 2000, 267, 3859-3868

[75] Arnau, J; Jørsensen, F; Madsen, SM; Vrang, A; Israelsen, H. Cloning, expression, and characterization of *Lactococcus lactis pfl* gene, encoding pyruvate formate-lyase. *J. Bacteriol.* 1997, 179, 5884-5891

[76] Melchiorsen, CR; Jokumsen, KV; Villadsen, J; Johnsen, MG; Israelsen, H; Arnau, J. Synthesis and posttranslational regulation of pyruvate formate-lyase in *Lactococcus lactis*. *J. Bacteriol.* 2000, 182, 4783-4788

[77] Takahashi, S; Abbe, K; Yamada, T. Purification of pyruvate formate-lyase from *Streptococcus lactis* and its regulatory role properties. J. Bacteriol. 1982, 149, 1034-1040

[78] Arnau, J; Jørsensen, F; Madsen, SM; Vrang, A; Israelsen, H. Cloning of the *Lactococcus lactis adhE* gene, encoding a multifunctional alcohol dehydrogenase, by complementation of a fermentative mutant of *Escherichia coli*. *J. Bacteriol.* 1998, 180, 3049-3055

[79] Neves, AR; Ramos, A; Costa, H; van Swam, II; Hugenholtz, J; Kleerebezem, M; de Vos, WM; Santos, H. Effect of different NADH oxidase levels on glucose metabolism of *Lactococcus lactis*: kinetics of intracellular metabolite pools by in vivo NMR. *Appl. Environ. Microbiol.* 2002, 68, 6332-6342

[80] Koebmann, BJ; Andersen, HW; Solem, C; Jensen, PR. Experimental determination of control of glycolysis in *Lactococcus lactis*. *Antonie van Leeuwenhoek* 2002, 82, 237-248

[81] Neves, AR; Pool, WA; Kok, J; Kuipers, OP; Santos, H. Overview on sugar metabolism and its control in *Lactococcus lactis*- The input from in vivo NMR. *FEMS Microbiol. Rev.* 2005, 29, 531-554

[82] Llanos, RM; Harris, CJ; Hillier, AJ; Davidson, BE. Identification of a novel operon in *Lactococcus lactis* encoding three enzymes for lactic acid synthesis: phosphofructokinase, pyruvate kinase, and lactate dehydrogenase. *J. Bacteriol.* 1993, 175, 2541-2551

[83] Crow, VL; Prichard, GG. Fructose 1,6-diphospate-activated L-lactate dehydrogenase from *Streptococcus lactis*: kinetic properties and factors affecting activation. *J. Bacteriol.* 1977, 131, 82-91

[84] Collins, LB; Thomas, TD. Pyruvate kinase of *Streptococcus lactis*. *J. Bacteriol.* 1987, 169, 5887-5890

[85] Poncet, S; Mijakovic, I; Nessler, S; Gueguen-Chaignon, V; Chaptal, V; Galinier, A; Boel, G; Maze, A; Deutscher, J. HPr kinase/phosphorylase, a Walker motif A-containing bifunctional sensor enzyme controlling catabolite repression in Gram-positive bacteria. *Biochim. Biophys. Acta* 2004, 1697, 123-135

[86] Monedro, V; Kuipers, OP; Jamet, E; Deutscher, J. Regulatory functions of serine-46-phosphorylated HPr in *Lactococcus lactis*. *J. Bacteriol.* 2001, 183, 3391-3398

[87] Poolman, B; Bosman, B; Kiers, J; Konings, WN. Control of glycolysis by glyceraldehyde-3-phosphate dehydrogenase in *Streptococcus cremoris* and *Streptococcus lactis* MG1363. *J. Bacteriol.* 2003, 185, 1564-1571

[88] Brøndsted, L; Hammer, K. Use of integration elements encoded by the temperate lactococcal bacteriophage TP901-1 to obtain chromosomal single-copy transcriptional fusions in *Lactococcus lactis*. *Appl. Environ. Micobiol.* 1999, 65, 752-758

[89] Andersen HW; Pedersen, MB; Hammer, K; Jensen, PR. Lactate dehydrogenase has no control on lactate production but has a strong negative control on formate production in *Lactococcus lactis*. *Eur. J. Biochem.* 2001, 268, 6379-6389

[90] Papagianni, M; Avramidis, N; Filiousis, G. Investigating the relationship between the specific glucose uptake rate and nisin production in aerobic batch and fed-batch glucostat cultures of *Lactococcus lactis*. *Enzyme Microb. Technol.* 2007, 40, 1557-1563.

[91] Gaudu, P; Lamberet, G; Poncet, S; Gruss, A. CcpA regulation of aerobic and respiration growth of *Lactococcus lactis*. *Mol. Microbiol.* 2003, 50, 183-192

[92] Palmfeldt, J; Paee, M; Hahn-Hägerdal, B; van Neil, EWJ. The pool of ADP and ATP regulates anaerobic product formation in resting cells of *Lactococcus lactis*. *Appl. Environ. Microbiol.* 2004, 70, 5477-5484

In: Glycolysis: Regulation, Processes and Diseases
Editor: Paul N. Lithaw
ISBN: 978-1-60741-103-1
© 2009 Nova Science Publishers, Inc.

Chapter II

The Cancer-Hypoxia/Decreased Respiration-Glycolysis Connection: New Insights from Nobel Prize-winner, Otto Warburg, MD, PhD

Brian Scott Peskin*

Chief Research Scientist
Cambridge International Institute for Medical Science,
Houston, Texas 77256, USA

Abstract

Everyone of true conscience must admit that over the last 30 years insufficient progress has been made in the "war to cure cancer." Otto Warburg, M.D., Ph.D., showed decades ago that development of cancer had a singular, prime cause. Each and every time cells (and tissues) were deprived of oxygen for a sufficient period of time, cancer developed. Furthermore, he clearly showed that the distinguishing feature of all cancer cells is the increase of anaerobic glycolysis and concurrent decrease of respiration—not merely excessive cell divisions. The significant increase in glycolysis observed in tumors has been verified today, yet few oncologists or cancer researchers understand the full scope of Warburg's work and its great importance. Without the use of Warburg's seminal discovery, cancer can never be truly cured—merely treated—although ineffectively, because when cancer returns from "remission," as is often the case, the patient has a high probability of death; treatments are ineffective. Extensive references to Warburg's original research are given.

* e-mail: prof-nutrition@sbcglobal.net, www.CambridgeMedScience.org

Introduction

> Any intelligent fool can make things bigger and more complex. It takes a lot of genius and a lot of courage to move in the opposite direction.
>
> —Albert Einstein, 1879-1955

This paper is about the incredible discovery of Nobel Prize-winner Otto Warburg, M.D., Ph.D., regarding cancer's *prime* cause—chronic systemic, cellular hypoxia (lack of oxygen), and cancer's *prime* characteristic—the ratio of respiration to fermentation (anaerobic glycolysis).

It is important to understand that Dr. Warburg always used actual *real-life* results as the basis of any scientific theory or explanation, *allowing the theory to fit the facts.* Unfortunately, this rarely happens with today's cancer researchers. They have it backwards—attempting to force the facts to fit their genetically based theories when their misguided theories do not fit the facts.

Significant glycolytic activity is a fundamental property of any tumor cell. Few oncologists today are familiar with Dr. Warburg's seminal work in this area; not surprisingly, progress both in preventing cancer and making significant improvements in treating the disease are lacking. Given the trillions of dollars spent on cancer research over the last 4 decades, there has been surprisingly little accomplishment compared to the great strides made, for comparable dollars spent, in other fields such as microelectronics and medical imaging technology.

Without understanding cancer's direct relationship with anaerobic glycolysis, I fear oncological treatments will continue to fall short of maximum effectiveness.

I am choosing to extensively reference Warburg's seminal work, specifically "The Metabolism of Tumours: Investigations from the Kaiser Wilhelm Institute for Biology"[1].

Glycolysis and Respiration

Throughout this paper we will use the term "glycolysis" to mean anaerobic (without oxygen) glycolysis with the end product of lactic acid. In humans, energy can be gleaned in two ways: through glycolysis or through cellular respiration. Glycolysis is the first step of each, although glycolysis does not require oxygen in any step of its chemical reactions.

When sufficient cellular oxygen is both plentiful and can be utilized, glucose is oxidized to pyruvate, which then enters the Krebs cycle. With insufficient cellular oxygen, the dominant glycolytic product is lactate. This latter process is known as anaerobic glycolysis.

Energy generation from stearic acid, the most commonly found fatty acid in triglycerides in the human body, can only occur in the mitochondria[2]. However, mitochondria can also beta-oxidize fatty acid molecules to form 2-carbon segments of acetyl-coenzyme A (acetyl-CoA) molecules, and the entire fatty acid molecule is broken down in this fashion. From each acetyl-CoA molecule split from a fatty acid, a total of 4 hydrogen atoms are released, and these are ultimately oxidized in the mitochondria to form large amounts of ATP—146 molecules from each molecule of stearic acid. This chapter will *not* focus on this pathway; cancer cares little about it.

Glycolysis occurs in the cytoplasm of the cell, not in a specialized organelle, such as the mitochondrion, and is the *one common metabolic pathway found in all living things*. Glycolysis is simply the splitting of glucose into 2 molecules of pyruvic acid; it then proceeds via fermentation to produce 2 net molecules of ATP, along with waste products. There are many types of fermentation but we will only concern ourselves with lactic acid fermentation because this is applicable to humans and cancer metabolism.

Cellular respiration (with oxygen) does not produce lactic acid; the pyruvate is completely broken down to CO_2 and H_2O, with vastly more energy cogeneration than glycolysis. Three molecules of O_2 are required for each molecule of pyruvic acid, and the end of cellular respiration produces a net 36 molecules of ATP per molecule of glucose after processing in the Krebs cycle and the electron transport chain. Cellular respiration may also be termed *aerobic* glycolysis.

In 1910, Dr. Warburg published the following: (1) "The *most important and completely unexpected result* of the present investigation is the proof that the plasma-membrane *as such* and *not because substances pass in or out* through it, plays an *important role* in the oxidative metabolism of the cell. (2) In section II this was *proved unquestionably*" (emphasis added).[1]

Dr. Warburg's discovery shows that it is the cell membrane itself that is key to proper physiologic functioning. Each of us has approximately 100 trillion cells, each containing a (bi)lipid membrane. As Dr. Warburg states, this important result—the membrane itself—was "completely unexpected."

Dr. Warburg proved decades ago in Germany, and it was confirmed by researchers in the United States, that when hypoxia—systemic oxygen deprivation—with 35% less cellular oxygen transfer occurs for a sufficient amount of time, cancer results.

Who Was Nobel Prize Winner Otto Warburg?

Dr. Warburg earned his doctorate in chemistry at the Berlin University in 1906 after initially studying under the great chemist, Emil Fisher. Warburg then studied medicine and earned his Doctor of Medicine at Heidelberg University in 1911.

How Significant is Otto Warburg?

We may gather some idea of the importance of Dr. Warburg's work by what his colleagues said of him. In 1931, Dr. Warburg was awarded the Nobel Prize in "Physiology or Medicine." Professor E. Hammarsten of the Royal Caroline Institute, a member of the Nobel Committee, said this to Dr. Warburg in his presentation speech: "Your bold ideas, but above

---

1 Although this experiment was performed with developing sea-urchin eggs and the "plasma-membrane" likely referred to what is termed the "fertilization-membrane," the insight is extraordinary.

all, your keen intelligence and *rare perfection in the art of exact measurement* have won for you exceptional successes, and *for the science of biology some of its most valuable material.*" In 1966, Dean Burk at the American National Cancer Institute said of Otto Warburg: "His main interests are Chemistry and *Physics of Life. In both fields no scientist has been more successful.*"

Chronology of Tumor and Cancer Discoveries

- The metabolism of tumors (1923-1925)
- The chemical constituent of the oxygen transferring respiratory ferment
- Origin of cancer cells (1956)
- Production of cancer metabolism in normal cells grown in tissue culture (1957-1968)
- Facultative anaerobiosis of cancer cells (1962-1965)
- Prime cause and prevention of cancer (1966-1969).

Dr. Warburg was one of the first cancer researchers. His insights and discoveries were incredible. Uniquely, despite his early successes and honors, Dr. Warburg continued to make major fundamental discoveries throughout his later years as well, capping off an amazingly fruitful 60-year career in research.

What Is Cancer?

While discussing the evils of cancer with a colleague, I realized how to simply explain what cancer really is. First let me state what it is *not*. It is *not an invader* in our bodies like a viral or bacterial infection. It is *not a genetic distortion* determined to kill us. It is *not an evil genius malcontent* buried deep within us waiting to strike its unsuspecting host. Cancer is none of these things.

Cancer is the body, at the cellular level, attempting to survive by reverting to a primitive survival mechanism. Surprisingly, it's that simple.

Hypoxia = Cancer

Over 80 years ago, Dr. Warburg proved that a 35% reduction in oxygen caused any cell to either die or turn cancerous. An amazing experiment by the Americans Goldblatt and Cameron in 1953 further confirmed this cancer/hypoxia connection[3], which was described by Warburg thus: "…[Goldblatt, an M.D. and Cameron] exposed heart fibroblasts in tissue culture to intermittent oxygen deficiency for long periods and finally obtained transplantable cancer cells. In the control cultures that they maintained without any oxygen deficiency, no cancer cells resulted"[4].

This experiment was conducted over a 2½-year time frame. The results were meticulously tabulated, and the conclusions rock solid. Dr. Warburg's work was extensively referenced in these scientists' paper, since his findings were very well known at that time.

Significantly, Goldblatt and Cameron also verified Dr. Warburg's finding (published in 1925)[5] *that a "respiration-impacted," destined-to-become cancerous cell could be stopped if it was oxygenated early enough.* In Goldblatt and Cameron's paper (p.527), it was reported:

> ...The length and frequency of exposure of the different [normal] cultures to nitrogen [cutting off oxygen] were varied greatly at first, in order to determine the periods that would prove *definitely injurious* in greater or less degree, but from which most of the *cultures recovered readily after the return to aerobic* [oxygenated] *conditions* were 15 minutes of nitrogen twice in 24 hours, for 3 successive days with an interval of 11¾ hours between successive exposures.
>
> It was found that even after exposure to nitrogen for ½ hour, 3 times in every 24 hours, for 7 consecutive days, with an interval of 7½ hours between successive exposures, recovery could still occur, although the injury was great; *but recovery was slower and less certain* after such long periods of anaerobiosis [oxygen deprivation]; and *some of the cultures did not recover.* (Emphasis added.)

The authors also noted that once damage was too great to the cell, then no amount of oxygen would return the cell's respiration back to normal—it was forever doomed to a cancerous life.

In 1955, two other American scientists and physicians, Malmgren and Flanigan, again confirmed these findings, publishing them in the medical journal, *Cancer Research*[6]. An especially clever and convincing experiment added to the long list of experiments clearly demonstrating that *oxygen deficiency is always present when cancer develops.* These physicians referenced Dr. Warburg's work on p. 478 of their publication.

Dr. Warburg analogized Malmgren and Flanigan's results with the development of cancer cells in his *Prime Cause and Prevention of Cancer* lecture[7]:

> If one injects tetanus spores, which can germinate only at very low oxygen pressures, into the blood of healthy mice, the mice do not sicken with tetanus, because the spores find no place in the normal body where the oxygen pressure is sufficiently low. Likewise, pregnant mice do not sicken when injected with the tetanus spores, because also in the growing embryo no region exists where the oxygen pressure is sufficiently low to permit spore germination. *However, if one injects tetanus spores into the blood of tumor-bearing mice, the mice sicken with tetanus, because the oxygen pressure* in the tumors can be so low that the spores can germinate. *These experiments demonstrate in a unique way the anaerobiosis* [low oxygen] *of cancer cells and the non-anaerobiosis* [normal oxygen] *of normal cells*, in particular the non-anaerobiosis of growing embryos.

Note: Rats and mice have much shorter lives than humans, so results, both good and bad, occur much faster, making them very useful in medical experiments.

We will focus more on the extensive use of physiology than of biochemistry in the cancer/glycolysis analysis.

The Metabolism of Cancer Cells

Dr. Warburg's ground-breaking paper, titled "The Metabolism of Carcinoma Cells," was published in the United States in *The Journal of Cancer Research* in 1925[5]. The paper was delivered as an address to the Rockefeller Institute in 1924, and much of this information had already been published in Germany in 1923.

Here are some of Dr. Warburg's glycolytic cancer findings that all oncologists and cancer researchers should be aware of: "...The result was *not* what we had anticipated ... *glucose brought the respiration to a standstill....*" Here it should be noted that cancerous tumors prefer sugar above all other metabolic fuels, and sugar stopped normal respiration. This effect does not occur in normal cells. Further, Warburg said, "In general it has been found that only tissue with unimpaired glycolytic activity is an integral property of the tumor cell. The conclusion drawn from this is that *glycolytic activity is an integral property of the tumor cell.*" Here, Dr. Warburg defined the fundamental property of any cancer tumor: its respiration is highly compromised. Finally, Warburg noted, "...The ratio splitting metabolism-oxidation metabolism for benign tumors is, however, displaced a long way in the direction of the oxidative metabolism. Malignant tumors produce three to four times more lactic acid per molecule of oxygen consumed than do benign tumors."

Here we are given tremendous insight into the difference between benign and cancerous tumors, and a key analytical tool to easily measuring the degree of malignancy.

Otto Warburg's Research

Dr. Warburg didn't play language games or use weasel words in reporting his results. He stated his findings definitively, based on extremely thorough and meticulous experimentation. Because he rarely used qualifying words to describe his findings, his anticancer discoveries and results offer sharp, definitive conclusions. He spent almost 60 years investigating cancer and he repeated experiments as many as 100 times before publishing. He did not draw conclusions lightly and he did not publish them until he was sure—which is why he was able to state them in definite terms. In contrast to the irresponsible tone so prevalent today, Dr. Warburg always held himself accountable for what he published. With Warburg's work, there was no need for the ubiquitous "new research shows..." that the old research was wrong and in need of correction. That is also why virtually *nothing he published was ever shown to be wrong later*—it was not just that he was sure, but that his conclusions were based upon unassailable science consistently repeated around the world.

As mentioned earlier, Professor E. Hammarsten of the Nobel Committee, in presenting Dr. Warburg with his Nobel Prize in 1931, made reference to Dr. Warburg's "*...rare perfection in the art of exact measurement...*" People may not have always agreed with his findings, but if they disagreed, they had no methodological basis for their disagreement.

Dr. Warburg even warned us decades ago that the cause of cancer would not be found in genetics—that research in this area would waste precious time and allow many more needless cancer mortalities..

Cancer is not Genetic

In his 1998 book, "*One Renegade Cell: How Cancer Begins*," author Dr. Weinberg presents an excellent summary, much of it quite technical, of the previous few decades of "advancement" in the fight against cancer[8]. The author is a professor of biology at MIT and former director of the Oncology Research Laboratory at the Whitehead Institute in Cambridge, Massachusetts.

The problem with modern cancer researchers' utter failure to find the *prime* cause of cancer or a valid means of preventing either the initial inception of the disease or a recurrence after remission has been their gradual shift from concentration on practical research to exploring academic and theoretical questions. Many of today's cancer researchers seem to live in a dream world where pet theories may be explored for years without leading to any real solutions to disease. Regarding the huge effort to explain cancer with genetics, Dr. Weinberg stated, "...Something was very wrong. The notion that a cancer developed through the successive activation of a series of oncogenes had lost its link to reality"[8].

Dr. Weinberg exposes and details failure after failure of cancer researchers to find cancer's cause or cure. More to the point, Dr. Weinberg states on page 67 that *cancer-causing "genes" are recessive—not dominant as everyone assumed!* On page 90 he reveals that "[F]ewer than one DNA base in a million appears to have been miscopied." Thus, the prime cause of cancer is *not* a genetic mutation. On page 95, Dr. Weinberg shares his opinion that the genetic discoveries made thus far are "sterile"—the prime cause of cancer is not "genetic."

On page 153, in the section, "Conquering Cancer by Preventing It," Dr. Weinberg states "We must *address the ultimate roots of cancer* before we make substantial reductions in cancer incidence. *Genes and proteins will help us little here*." How much clearer can Dr. Weinberg make it that cancer is not genetically based?

Weinberg clearly makes the point that all the modern research roads over the past 30 years geared toward finding the cause of cancer have led nowhere.

The Genetic Fallacy is Exposed Again—Internationally

The following article was published internationally via the excellent Internet publication *Medical News Today,* in an article titled "Cancer Comes Full Circle"[9], which refers to an article published in *Nature*[10]:

> 'This study demonstrates how structure and function in a tissue are intimately related, and how loss of structure could itself lead to cancer,' says Mina Bissell, who pioneered the view that *a cell's environment is as important as its genes* in determining the formation and progression of tumors.
>
> ...But a number of investigators, including Bissell and her colleagues, have shown that the genetic alterations of oncogenes are not, as once believed, sufficient in themselves to cause cancer. Even activated oncogenes require changes in the tissue structure to produce cancer. (Emphasis added.)

Herein lies the cry to look elsewhere than genetics for the roots of cancer, and we reiterate that Dr. Warburg has already given us that key: glycolysis. Once again, a cry to look at the tissue structure is made. Tissue physiology can show us that glycolysis rather than respiration dominates in cancer.

Confirmation that Cancer Increases with Lack of Oxygen

An article in the cancer medical journal, *Radiotherapy and Oncology*, makes Dr. Warburg's #1 fact clear[11]:

> After a median follow-up of 19 months (range 5-31 months), Kaplan-Meier-life table analysis showed *significantly lower survival and recurrence-free survival* for patients with a median pO2 of ≤ 10 mm Hg *compared to those with better oxygenated tumors* (median pO2 > 10 mmHg).
>
> The Cox proportional hazards model revealed that the *median pO2* and the clinical stage according to the FIGO are *independent, highly significant predictors of survival and recurrence-free survival.*
>
> We conclude from these preliminary results that *tumor oxygenation* as determined with this standardized procedure appears to be a new independent prognostic factor influencing survival in advanced cancer of the uterine cervix. (Emphasis added.)

Benign versus Cancerous Tumors

What differentiates a cancerous tumor from a non-cancerous (benign) tumor? The cells of both tumors demonstrate essentially the same "mindlessness"—lost cellular intelligence. It's all a matter of degree of respiration impairment. Dr. Warburg had already verified and published this fact in 1925 in *The Journal of Cancer Research*[5]. Dr. Warburg's paper makes it quite clear:

> Thus the quantitative difference between malignant and benign tumors becomes a qualitative one, when we pass from benign tumors to normal growth. The respiration of normally growing tissues suffices to bring about the disappearance of the glycolysis-products, whereas in tumors the *respiration is too small for this. This, then, is the difference between ordered and disordered growth.*
>
> …From the embryonal type of metabolism there has again arisen the tumor type—benign or malignant, *depending on the duration of the oxygen deficit.*
>
> In this manner [adding higher degrees of cyanide to curtail respiration] we obtain from the embryonic type of metabolism the tumor type—the benign tumor type when the concentration of cyanide is low [less impacted respiration]; the malignant type, when it is high [highly impacted cellular respiration].… [T]here has again arisen the tumor type—benign or malignant, depending upon the duration of the oxygen deficit.

Dr. Warburg's genius was unprecedented in making these seminal discoveries regarding the metabolism of cancer.

Dr. Warburg clarifies this in his own words:

> The most important fact in this field is that *there is no physical or chemical agent with which the fermentation of cells in the body can be increased directly*; for increasing fermentation, a long time and many cell divisions are always necessary. The *mysterious latency period* of the production of cancer is, therefore, nothing more than the time in which the *fermentation increases after a damaging of the respiration*. This time differs in various animals; *it is especially long in man and here often amounts to several decades,* as can be determined in the cases in which the time of the respiratory damage is known—for example, in arsenic cancer and irradiation cancer. (Emphasis added.)

Warburg makes the startling statement that you cannot make a cell increase its fermenting capability unless lack of oxygen is at the root of the process.

Another landmark Warburg paper titled, "The Metabolism of Tumors in the Body,"[12] published by The Rockefeller Institute for Medical Research in New York in 1928, provides additional insight by stating *all tumors so far tested behave fundamentally alike*. Although this statement was already published in the *Journal of Cancer Research* paper three years before in 1925, it is worthy of repeating this important fact. Further, the authors state, "The *tumor cell is more versatile than the normal cell* as far as the obtaining of energy is concerned. *It can choose between fermentation and respiration*, while the normal cell is confined to respiration." This makes cancer cells much harder to kill than normal cells, and explains why prevention is so important, so that cancer never has a chance to start to develop.

A top biochemistry textbook in use in 1979 at MIT, where I matriculated, discussed the decreased aerobic (respiration)/increased anaerobic (glycolysis) relationship found with cancer cells[13]. On page 849 it states, "The rate of oxygen consumption of cancer cells is somewhat below the values given by normal cells. However, *malignant cells tend to utilize anywhere from 5 to 10 times as much glucose as normal tissues* and *convert most of it into lactate*...." Note that more glucose is required because of the lack of oxygen utilization for energy.

Lactic Acid Burn: A "Do-It-Yourself" Test

If you have ever worked out with weights, then you have likely already experienced "lactic acid burn." It is a burning sensation that comes from lactic acid buildup in your muscles, produced when they *ferment glucose for energy*—much in the same way that a cancer cell does. "Lactic acid burn" becomes a *problem of the past* when cellular oxygenation is increased.

Do lactic acid levels really increase in the blood if you have cancer? Yes. This fact was published back in 1925 by Dr. Warburg in his cancer journal article[5]. Current researchers also confirm the increase in lactic acid.

In investigating the relationship between lactate levels and human cervical cancer, Walenta et al.[14] found that the *metastatic spread* of uterine cervix carcinomas and neck cancers were closely correlated with lactate concentration in the primary lesion. However, the authors also noted that lactate concentrations were *significantly higher* (p = .001) in tumors *that had spread metastatically* (mean ± SD, 10.0 ± 2.9 micromol/g; n = 20) compared to malignancies in patients in which metastases *did not occur* (6.3 ± 2.8 micromole/g; n = 14). Furthermore, the survival probabilities of patients that had low tumor lactate values were *significantly higher* compared to patients with high tumor lactate concentrations. The authors concluded, "Tumor lactate content may be used as a prognostic parameter in the clinic. Furthermore, these findings are in accordance with data from the literature showing the *presence of hypoxia in cervical tumors is associated with a poorer survival rate*." (Emphasis added.)

In discussing aerobic glycolysis, Lu et al.[15] note that all cancer cells display *high rates of aerobic glycolysis*, a phenomenon historically known as the Warburg effect, but add "the relevance of the Warburg effect to cancer cell biology has remained obscure." In their study, they discovered that the ability of *glucose to stimulate* HIF-1 (hypoxia-inducible Factor 1), a factor which *increased with lack of oxygen in cancer*, increased in parallel with the growth of the cancer, tumor development, tumor angiogenesis, and poor prognosis. Moreover, this effect could not be mimicked by using the glucose analog, 2-deoxyglucose, suggesting that the *metabolism of glucose was required for this effect to occur.*

This 2002 cancer study makes it clear that Dr. Warburg is still known—*cancer's high glycolysis is termed the "Warburg effect"—but cancer researchers still have no idea how to make use of his discovery practically*, as evidenced by the second point above. These researchers found that HIF-1 responds directly to low oxygen, and is *stimulated* by glucose. In 2002, the Department of Biochemistry and Molecular Biophysics at the University of Arizona stated that the "aerobic glycolysis phenotype," first described by Otto Warburg in 1924, might be central to the process of carcinogenesis itself.

Schwickert et al.[16] also stated the same result: the higher the lactic acid the greater the spreading of the cancer. An Italian study[17] once again stated the same result, with those patients having *higher lactic acid levels also having the highest rates of cancer recurrence after treatment.* Druml et al.[18] also discussed how the *increased lactic acid comes from the leukemic cells themselves* and no other cause or other source. The fact that cancer causes an increase in lactic acid is well known.

Researchers from the Harvard Medical School and Massachusetts General Hospital's Department of Radiation Oncology also showed that low pH always comes from lactic acid as well as other by-products[19]. Their research confirms the above results.

I want to make this confirmation quite clear, so there is no misinterpretation. *Increased lactic acid output from cancer cells can always be used as a cancer marker.*

The bottom line is to keep cellular oxygenation levels high. In this fashion, as Dr. Warburg so clearly discovered, cancer can never start. Lactic acid levels naturally remain low when you are cancer-free and rise consistently, depending on how aggressive the cancer becomes. Although lactic acid levels increase primarily in the tumor tissue itself, lactic acid levels rise in the blood, too—and it is easy to have it measured.

Is Anaerobic Glycolysis (Running on Sugar) Really Significant for a Cancerous Cell?

Yes, it is. There is a drastic difference between cancerous and non-cancerous cells, and this difference is the greatest such difference. Dr. Warburg stated on p.151 of *The Metabolism of Tumours*:

> Blood forms per hour a quantity of lactic acid equivalent to 0.1% of its dry weight, as compared with 12.4% formed by the tumour. The glycolytic action of the carcinoma tissue is *124 times greater* than the glycolytic action of *blood*.... Hence, carcinoma tissue forms *200 times as much lactic acid* as a resting frog's muscle and *8 times as much* lactic acid as a working frog's muscle working at maximum normal capacity[1]. (Emphasis added.)

As you can see, there is a significantly more lactic acid formation from cancerous tissue. Even muscle, which uses sugar as its prime fuel under intense physical exertion, *still produces a significant eight times less lactic acid than cancerous tissue.*

Furthermore, it is an easy experiment to show that all animal cells can run without oxygen to a certain extent (although not efficiently). However, with oxygen, most animal cells do not use glycolysis, as Warburg states on p. 60:

> The first case occurs when we are working under anaerobic conditions. Any animal cells serve as experimental material, since all animal cells glycolyse [utilize significant sugar as prime metabolic fuel] under anaerobic conditions.... The second case arises when we are working under aerobic conditions with cells which do not glycolyse aerobically. This is the case with most normal animal tissues.

Louis Pasteur was the father of the field of stereochemistry, the "pasteurization" process, the "germ theory" (explaining the cause of most infectious disease)—one of the most important discoveries in medical science—a pioneer in the treatment of rabies, and founder of a revolution in verifiable science by demanding, "*Do not put forward anything that you cannot prove by experimentation.*" His work became the foundation for the science of microbiology and a cornerstone of modern medicine. *He understood the connection between cancer and —lack of respiration— and it was well known in 1861.* Dr. Warburg[20] described the significance of one of Pasteur's discoveries[21]:

> As is well known, Pasteur found that respiration 'inhibits' fermentation. If he placed cells, which fermented under anaerobic conditions, in oxygen, the respiration, which now began caused either the diminution or the disappearance of the fermentation. Respiration and fermentation are thus connected by a chemical reaction I call the 'Pasteur reaction' after its discoverer.
>
> It is not so much the inhibition of fermentation itself, which is characteristic of the Pasteur reaction, but rather the relationship between the inhibition of the fermentation and the respiration, i.e. the quotient: (anaerobic fermentation – aerobic fermentation) / respiration.

Warburg also commented that the quotient, which O. Meyerhof (another of Warburg's protégées and a Nobel Prize-winner) was the first to determine in the case for muscle, was purely an experimental quantity—i.e., it was based on real-life results—and independent of any theory. This means that there is no way to theoretically determine or guess that this result indeed occurs. *If a cancer scientist or researcher wasn't informed about this discovery, there is no way he or she would find it from other fields independently.*

Detailed Excerpts from Dr. Warburg

In this section, I provide cancer discoveries, taken from a speech Warburg gave at the 1966 Nobel-Laureates Conference in Lindau, Germany. The name of the address was "The *Prime* Cause and *Prevention* of Cancer"[7][2]:

> But, even for cancer, there is only one main cause. Summarized in a few words, the prime cause of cancer is the replacement of the respiration of oxygen in normal body cells by a fermentation of sugar.

Comment

Normal cell respiration is replaced by energy production through the fermentation of sugar. This means that carbohydrates are utilized as cancer's prime fuel instead of proteins or fats. Cancer cells grow from the fuel of carbohydrates. When a cell cannot get the oxygen it requires, then it turns to fermentation of sugar for its energy. Next, Warburg noted:

> Neither *genetic codes* of anaerobiosis nor *cancer viruses* are alternatives today, because no such codes and no such viruses in man have been discovered so far....

Comment

Dr. Warburg makes a very clear statement: cancer has *no* genetic basis and *no* viral basis that he could find. Nothing has changed since he made this statement over 40 years ago. But regardless of this groundbreaking insight, even today most medical institutions continue to look for the answers in the wrong areas. Thus, Warburg adds:

> Because no cancer cell exists, the respiration of which is intact, it cannot be disputed that *cancer could be prevented if the respiration of the body cells would be kept intact....*

[2] English Edition by Dean Burk, National Cancer Institute, Bethesda, Maryland, USA.

Comment

Here, Dr. Warburg makes it perfectly clear that there is *no* cancer cell whose cell respiration is intact; therefore, cancer should be preventable if cellular respiration can be kept intact.

It is important to note that these facts regarding cell respiration are fundamental and apply to *all* cancer cells. Dr. Warburg states that *no cancer cell exists that has fully intact oxygen respiration. All cancer cells share* this unique characteristic:

> If it is so much decreased that the oxygen transferring *enzymes are no longer saturated with oxygen*, respiration can decrease *irreversibly* and *normal cells* can be *transformed* into facultative *anaerobes.*

Comment

Once sufficient oxygen-deficiency damage is done to a cell, it *cannot ever be repaired.* There is a "point of no return." Therefore, Dr. Warburg's amazing finding implies that *cancer prevention is the key.* Damaged cells only become cancerous because they have fermentation to turn to instead of dying. Then, they live and multiply, spreading the cancer. Warburg adds:

> *All normal cells* meet their energy needs by the *respiration of oxygen*, whereas *cancer cells* meet their energy needs in great part by *fermentation.* From the standpoint of chemistry and physics of life this difference between normal and cancer cells is so *great* that *one can scarcely picture a greater difference. Oxygen gas*, the donor of energy in plants and animals is *dethroned in the cancer cells* and *replaced* by an energy yielding *reaction of the lowest forms*, namely a *fermentation of glucose.*

Comment

Cancer cells are so different from normal cells that a greater difference cannot be imagined. Oxygen gas is relegated to a lower importance in cancer cells. Cancer cells thrive on sugar—the fuel of fermentation. Thus, Warburg says:

> Cancer cells can actually grow in the body almost with only the energy of fermentation.
>
> In the mouse, if one provides an oxygen pressure so reduced that the *oxygen respiration is partially inhibited*, the purely aerobic metabolism of the mouse embryonic cells is quantitatively *altered in 48 hours in the course of two cell divisions.*

Comment

Researchers must be wary of animal studies; however, in this case a direct human analogy is appropriate. Dr. Warburg's amazing experiments showed how quickly cells could enter the highway to cancer (although it takes them a long time to become fully cancerous *in the human body*—often *several decades* compared to "test tube" experiments conducted outside the body [i.e., *in vitro*]). Although this occurred in a mouse, the analogy with humans is correct—human cancer cells require little energy to live and oxygen-deprived cells replicate quickly. On irreversibility, Warburg noted:

> If one increases to the original high oxygen pressure, and allows the cell to grow further, the cancer metabolism *remains*, it's an irreversible process.

Comment

It is of paramount importance to understand that cancer is an *irreversible transformation.* At all costs it must be prevented. Once you have a cancer cell, there is no returning it back to normal. It is impossible. That's why it is termed irreversible. On the cause of transformation, Warburg said:

> We find by experiment about *35% inhibition of oxygen respiration* already suffices to bring about such a transformation during cell growth.

Comment

Only about *one-third less* oxygen transfer to a cell causes *irreversible cancer cells to form*. For maximum anti-cancer support, we need to fully oxygenate our cells, so possible glycolytic action potential never fully occurs. Dr. Warburg makes this clear. This is the cellular analogy of being choked to death:

> In any case, during the *cancer development the oxygen-respiration always falls, fermentation appears*, and the highly differentiated cells are transformed to fermenting aerobes, which have lost all their body functions and retain only the now useless property of growth.... The meaning of life disappears.

Comment

Cancer *always* shows itself by non-availability of oxygen. A cancer cell lacks intelligence; it is a useless, "mindless, growing machine." On the chemistry, Warburg noted:

> The end-products of fermentation [the metabolic process associated with cancer] are reached by one single reaction. On the other hand, the end-products of the oxidation of pyruvic acid [the metabolic process of normal, healthy cells] are only reached after many additional reactions. Therefore, when cells are harmed, it is probable that first respiration is harmed.

Comment

Normal cell respiration is significantly more biochemically complicated than simple fermentation of sugar. Cancer cells prefer the easier fermentation route to live, in part, because there is no "intelligence" left in these cells to direct a complicated oxygen-breathing mechanism. The cancerous cell loses its ability to function in a normal, healthy way, because it has become dumb. Dr. Warburg points out that it is likely that the first harm to the cell that occurs is likely to be the harm to its respiration.

Dr. Warburg spoke eloquently at the German Central Committee for Cancer Control[4][3]:

> What was formerly only qualitative now becomes quantitative. What was formerly only probable has now become certain. The era in which the fermentation of the cancer cells or its importance could be disputed is over, and no one today can doubt that we understand the origin of cancer cells if we know how their large fermentation originates, or, to express it more fully, if we know how the damaged respiration and the excessive fermentation of the cancer cells originate.

Why Wasn't this Information Known Earlier?

Dr. Warburg continued:

> Moreover, during the first decades after 1923 glycolysis and anaerobiosis were *constantly confused, so that nobody knew what was specific for tumors*. The three famous and decisive discoveries of Dean Burk and colleagues of the National Cancer Institute at Bethesda (USA) were of the years 1941, 1956 and 1964: first, that the metabolism of the regenerating liver, which grows more rapidly than most tumors, is not cancer metabolism, but perfect aerobic embryonic metabolism; second, that cancer cells, descended in vitro from one single normal cell, were *in vivo* the more malignant, the higher the fermentation rate; third, that in vivo growing hepatomas, produced in vivo by different carcinogens, were in vivo the *more malignant, the higher the fermentation rate*. Furthermore, the very *unexpected and fundamental fact*, that tissue culture is carcinogenic and that a *too low oxygen pressure is the intrinsic cause* were discovered in the years 1927 to 1966. Anaerobiosis of cancer cells was an established fact only since 1960 when methods were developed to measure the oxygen pressure inside of tumors in the living body.
>
> At first, one would think that it is immaterial to the cells whether they obtain their energy from respiration or from fermentation, since the energy of both reactions is transformed into the energy of adenosine triphosphate, and yet adenosine triphosphate = adenosine triphosphate. This equation is certainly correct chemically and energetically, but it is *incorrect morphologically,* because, although *respiration takes place for the most part in the structure of the grana* [mitochondria], *the fermentation enzymes are found for a greater part in the fluid protoplasm*. The adenosine triphosphate synthesized *by respiration therefore involves more [cell] structure* than the adenosine triphosphate

---

[3] Otto Warburg was director of the Max Planck Institute for Cell Physiology, Berlin-Dahlem, Germany. This article is based on a lecture delivered at Stuttgart on May 25, 1955 before the German Central Committees for Cancer Control. Translation by Dean Burk, Jehu Hunter, and WH Everhardy at the National Institutes of Health (USA).

synthesized by fermentation. Thus, it is as if one reduced the same amount of silver on a photographic plate by the same amount of light, but in one case with diffused light and in the other with patterned light. In the first case, a diffuse blackening appears on the plate, but in the second case, a picture appears; however, the same thing happens chemically and energetically in both cases. Just as the one type of light energy involves more structure than the other type, *the adenosine triphosphate energy involves more structure when it is formed by respiration than it does when it is formed by fermentat*ion.

Comment

In other words, normal cell respiration takes place in the presence of a more differentiated cell structure; a cancer cell's fermentation involves less structure. Warburg continued:

> Moreover, *it was known for a long time* before the advent of crystallized fermentation enzymes and oxidative phosphorylation that *fermentation*—the energy-supplying reaction of the lower organisms—*is morphologically inferior to respiration. Not even yeast, which is one of the lowest forms of life, can maintain its structure permanently by fermentation alone; it degenerates to bizarre forms*. However, as Pasteur showed, it is rejuvenated in a wonderful manner, if it comes in contact with oxygen for a short time. "I should not be surprised," Pasteur said in 1876 in the description of these experiments, "if there should arise in the mind of an attentive hearer a presentiment about the causes of those great mysteries of life which we conceal under the words youth and age of cells." *Today, after 80 years, the explanation is as follows: the firmer connection of respiration with structure and the looser connection of fermentation with structure.*
>
> *This, therefore, is the physicochemical explanation of the de-differentiation of cancer cells***.** If the structure of yeast cannot be maintained by fermentation alone, one need not wonder that highly differentiated body cells lose their differentiation upon continuous replacement of their respiration with fermentation.
>
> Since the increase in fermentation in the development of cancer cells takes place gradually, there must be a *transitional phase between normal body cells and fully formed cancer cells*. Thus, for example, when fermentation has become so great that de-differentiation has commenced, but not so great that the respiratory defect has been fully compensated for energetically by fermentation, we may have cells which indeed *look like cancer cells but are still energetically insufficient*. Such cells, which are clinically not cancer cells, have lately been found, not only in the prostate, but also in the lungs, kidney, and stomach of elderly persons. *Such cells have been referred to as "sleeping cancer cells*."
>
> Cancer cells originate from normal body cells in *two phases*. The *first phase is the irreversible injuring of respiration*. Just as there are many remote causes of plague—heat, insects, rats—but only one common cause, the plague bacillus, there are a great many remote causes of cancer—tar, rays, arsenic, pressure, urethane—*but there is only one common cause into which all other causes of cancer merge, the irreversible injuring of respiration*.
>
> Physics cannot explain why the two kinds of energy [glycolysis vs. respiration] are not equivalent in differentiation; but chemistry may explain it. Biochemists know that the results of both respiration and fermentation are stored in ATP; indeed, the basic mechanism of ATP creation is the same, but the reactions used to generate ATP

molecules are *quite different*. If one applies this knowledge to carcinogenesis, it seems that only *oxidative phosphorylation but not fermentative phosphorylation can differentiate,* a result that may in future explain the mechanism of differentiation.

Yet biochemistry can explain already today why fermentation arises, when respiration decreases.... The *pathways* of respiration and fermentation *are common as far as pyruvic acid*. Then the pathways diverge. The endproducts of fermentation [are] reached by one single reaction, the reduction of pyruvic acid by dihydronicotinamide to lactic acid. On the other hand, the endproducts of the oxidation of pyruvic acid, H_2O and CO_2, are only reached after many additional reactions. Therefore, when cells are harmed, it is probable that first respiration is harmed.

In this way the frequency of cancer is explained by reasons of probability.

Conclusion: How Cancer Occurs and How to Stop Cellular Respiration Reverting to Glycolysis

Heart attacks can stem from lack of oxygen. As Warburg discovered, so does cancer. Most normal, healthy cells get most of their energy by using oxygen in a process called "respiration." This can be contrasted with the way cells utilize energy *without* sufficient oxygen, called "fermentation," or glycolysis. Fermentation of sugar provides a way for cells to keep going energetically even in the presence of partial oxygen deprivation. In the presence of oxygen deficiency, cells that cannot obtain enough energy through fermentation perish. But the cells that succeed in utilizing fermentation exhibit their innate will to survive; these are the ones that do not die from the oxygen deficiency.

As directed by nature, in using this alternative source of energy, our cells are fulfilling their primary mission, which is to stay alive and reproduce. This takes place on all levels for all living things, and in the case of oxygen deficiency, cells are struggling to survive in a hostile environment of humans' own making. That's right; we unknowingly have forced our own cells to become cancerous.

After it begins, the disease worsens, since most humans hosting it never feel the cancer growing and so we do not take corrective measures. Once it shows up in lab tests, you have been hosting the cancer cells for years.

Nature has given every cell the potential to survive without oxygen, through fermentation. If that potential is not developed enough, then the cell will die when the oxygen drops below the 35% threshold. If none of our cells could run without oxygen, they would die immediately with no possible chance of future survival. Chronic deficiency of oxygen damages the mitochondria (energy producers) of the cell, so the cell, if it can, reverts to the ancient energy source of fermentation of sugar. A cancer cell running on fermentation can stay alive (without growing) with just 20% of a normal cell's energy.

But one major problem is that this method is very inefficient. The cells that *can run on fermentation without oxygen* stay alive and become more prevalent as the other cells die. But there is a huge price to be paid: lack of cellular intelligence. Regarding cancer, anaerobic glycolysis leads to life but not to intelligence, and over time, if not stopped so that cellular respiration dominates, often death.

Can cellular oxygenation levels stay high so glycolysis never occurs? Yes. With today's science it is known how to guarantee maximum cellular oxygenation. This is the subject of a future paper.

References

[1] Warburg, O. The Metabolism of Tumours: Investigations from the Kaiser Wilhelm Institute for Biology. Translated by Dickens, F. Constable & Co Ltd., 1930, 56 (out of print). Ref: *Hoppe-Seyler's Zeitschr Physiol Chem,* 1910 66, 305.

[2] Guyton, A; Hall, J. Textbook of Medical Physiology. 9th ed. Philadelphia, PA: W.B. Saunders; 1996:868.

[3] Goldblatt, H; Cameron, C. Induced malignancy in cells from rat myocardium subjected to intermittent anaerobiosis during long propagation in bitro. *J. Exp. Med.* 1953 97, 525-552.

[4] Warburg, O. On the origin of cancer cells. *Science.* 1956 123, 309-314.

[5] Warburg, O. The metabolism of carcinoma cells. *J. Cancer Res.* 1925 9, 148-163.

[6] Malmgren, RA; Flanigan, CC. Localization of the vegetative form of Clostridum tetani in mouse tumors following intravenous spore administration. *Cancer Res.* 1955 15, 473-478.

[7] Warburg O. The prime cause and prevention of cancer (Lindau Lecture). Revised ed. Würzburg, Germany: Konrad Triltsch, 1969. Accessed August 11, 2006. Retrieved from: http://www.hopeforcancer.com/OxyPlus.htm.

[8] Weinberg, RA. One Renegade Cell: How Cancer Begins. New York: Basic Books; 1998.

[9] Cancer comes full circle. Medical News Today 2005. Accessed August 5, 2008. Available at: http://www.medicalnewstoday.com/medicalnews.php?newsid=27058#.

[10] Radisky, DC; Levy, DD; Littlepage, LE, et al. Rac1b and reactive oxygen species mediate MMP-3-induced EMT and genomic instability. *Nature.* 2005 436, 123-127.

[11] Hockel, M; Knoop, C; Schlenger, K, et al. Intratumoral P02 predicts survival in advanced cancer of the uterine cervix. *Radiother. Oncol.* 1993 26, 45-50.

[12] Warburg, O; Wind, F; Negelein, E. The metabolism of tumors in the body. *J. General Physiol.* 1927 8, 519-530.

[13] Lehninger, AL. Biochemistry. New York: Worth Publishers; 1976.

[14] Walenta, S; Wetterling, M; Lehrke M, et al. High lactate levels predict likelihood of metastases, tumor recurrence, and restricted patient survival in human cervical cancers. *Cancer Res.* 2000 60, 916-921.

[15] Lu, H; Forbes, RA; Verma A. Hypoxia-inducible factor 1 activation by aerobic glycolysis implicates the Warburg effect in carcinogenesis. *J. Biol. Chem.* 2002 277, 23111-23115.

[16] Schwickert, G; Walenta, S; Sundfør, K; Rofstad, EK; Mueller-Klieser, W. Correlation of high lactate levels in human cervical cancer with incidence of metastasis. *Cancer Res.* 1995 55, 4757-4759.

[17] Bacci, G; Capanna, R; Orlandi, M, et al. Prognostic significance of serum lactic acid dehydrogenase in Ewing's tumor of bone. *Ric. Clin. Lab.* 1985 15, 89-96.
[18] Druml, W; Kleinberger, G; Neumann, E; Pichler, M; Gassner, A. [Acute leukemia associated with lactic acidosis] [article in German]. *Schweiz Med. Wochenschr.* 1981 111, 146-150.
[19] Helmlinger, G; Sckell, A; Dellian, M; Forbes, NS; Jain, RK. *Clin. Cancer Res.* 2002 8, 1284-1291.
[20] Warburg, O; Posener, K; Negelein E. The metabolism of cancer cells. *Biochem. Zeitschr.* 1924 152, 309.
[21] Pasteur L. [The influence of oxygen on the development of yeast and alcoholic fermentation] [article in French]. *Bull. Soc. Chim. Paris.* 1861, 79-80.

In: Glycolysis: Regulation, Processes and Diseases
Editor: Paul N. Lithaw
ISBN: 978-1-60741-103-1
© 2009 Nova Science Publishers, Inc.

Chapter III

Pattern Formation and Dissipation in a Model Glycolytic System: The Effect of Complexing Reaction with the Activator

Arun K. Dutt *

Faculty of Computing, Engineering and Mathematical Sciences (Du Pont Building), University of the West of England (Frenchay Campus), Bristol BS16 1QY, UK

Abstract

The effect of complexing reaction of the activator ADP has been investigated in a model glycolytic reaction-diffusion system generating Hopf and Turing wave instabilities. The complex formation reaction with the activator species reduces drastically the domain of homogeneous Hopf bifurcation in the parameter space producing more Turing region, where Turing bifurcation may be initiated by inducing inhomogeneous perturbations. Our numerical results conform to the expectation that Hopf wavelengths depend strongly on the degree of complexing reaction of the activator, whereas Turing wavelengths don't. For this model system, the reaction velocity and entropy production as a function of the reaction affinity are computed and the results interpreted in terms of the efficiency of biochemical engines.

1. Introduction

Oscillations in biochemical systems has become an area of active research in recent years. Degn, Olsen and coworkers[1-4] made investigations on the experiments and computer simulations of sustained and complex oscillations in the peroxidase-oxidase reaction[5,6]

* Present Address : 16 Ghanarajpur(Jalapara), Post: Dhaniakhali, Dist.Hooghly, WB 712302, India ; Email : dutt_arun@yahoo.com

involving the oxidation of NADH by molecular oxygen mediated by horseradish peroxidase enzyme. Subsequently, the experiments and numerical investigations of the Olsen and Degn model were carried out by Larter et al.[7] – it produces an entire sequence of complex and chaotic oscillations.

Self-oscillations in glycolysis has been discovered in intact yeast cells and yeast cell extracts, and also in beef-heart extracts[8]. The phosphofructokinase (PFK1) reaction is considered as being the possible source of self-oscillations in glycolysis.

The mechanistic features[9-12] of the glycolytic oscillations in yeast are – (a) a steady input of substrates such as glucose or fructose maintaining the system far from thermodynamic equilibrium ; (b) the allosteric properties of the enzyme PFK1 providing the basis of oscillations ; (c) a feedback effect on the activity of PFK1 through the adenosine nucleotide system - the allosteric regulation of PFK1 activity by ATP and AMP ; and (d) a sink for the products of the PFK1 reaction through pyruvate kinase coupled to the adenosine nucleotide system.

This paper is organized in the following way. In Section 2, we have formulated the reversible Sel'kov model, a mathematical model of glycolytic oscillations - by the application of law of mass action and molecular diffusion, we have derived the partial differential equations in dimensionless form. In Section 3, the partial differential equations have been modified to include the presence of complexing reaction with the activator species. In Section 4, analytical calculations have been made discussing Hopf and Turing bifurcations. In Section 5, we have presented a nonequilibrium thermodynamic description of this model system in terms of the rate of chemical entropy production (e.p.). Lastly in Section 6, numerical results have been presented and efforts have been made to correlate the results with the experimental data from literature.

2.The Model

Sel'kov [13] has proposed a simple kinetic model (I) of the enzyme catalysis with product activation of the enzyme, which exhibits limit cycle oscillations. Here, the substrate S (ATP) supplied by a certain source at a constant rate (v_1) is converted irreversibly into a product P (ADP). The product P (ADP) is removed by an irreversible sink at another constant rate (v_2.). The free enzyme E (PFK1) is inactive by itself, but becomes active after combination with γ product molecules to form the complex EP^{γ}.

$$\begin{aligned}
&\text{Source} \xrightarrow{v1} S + EP^{\gamma} \leftrightarrows SEP^{\gamma}, \\
&SEP^{\gamma} \rightarrow EP^{\gamma} + P \xrightarrow{v2} \text{Sink}, \qquad \text{(I)} \\
&\gamma P + E = EP^{\gamma}.
\end{aligned}$$

Based on Sel'kov's model, Richter et al.[14] have proposed the following three reversible kinetic steps, known as the reversible Selkov model (II),

$$A \leftrightarrows S \qquad (2.1)$$

$$S + 2P \leftrightarrows 3P \qquad (2.2) \qquad (II)$$
$$P \leftrightarrows B \qquad (2.3)$$

where A and B are controllable source and sink concentrations respectively — $k_{\pm i}$ (i = 1, 2, 3) are, respectively, the rate constants of the three steps (+ and – subscripts are, respectively, for forward and reverse reactions). It is worthwhile to note that the reversible Sel'kov model (II) steps account for only the oscillatory part of the mechanism, not the complete glycolytic mechanism itself, which is rather complicated. When described in terms of activator/inhibitor model, the autocatalyst P is the activator and S, the inhibitor, the R-D equations [15-17] in one dimension are given by:

$$\partial P/\partial t = k_2SP^2 - k_{-2}P^3 - k_3P + k_{-3}B + D_p(\partial^2P/\partial x^2)$$
$$\partial S/\partial t = k_1A - k_{-1}S - k_2SP^2 + k_{-2}P^3 + D_s(\partial^2S/\partial x^2) \qquad (2.4)$$

where P and S represent the concentrations of the respective species, D's are diffusion coefficients and x is the geometrical coordinate. Since the specific rate constants of the model steps are not known, it is necessary to scale the reaction part[14,16] and diffusion part[18] appropriately as given below in equations (2.5) and (2.6) respectively,

$$k_3t = \tau \ ; S/N = s \ ; \quad P/N = p \ ; N = (k_3/k_2)^{1/2} \ ;$$
$$a = (k_1A)/(k_3N) \ ; \quad b = (k_{-3}B)/(k_3N) \ ; \quad \kappa = (k_{-1}/k_3) \ ;$$
$$K_2 = (k_{-2}/k_2) \qquad (2.5)$$
$$\rho = x\,(k_3/D_p)^{1/2} \qquad (2.6)$$

Therefore, the equations (2.4) take the form as given below.

$$\partial p/\partial \tau = -p + b + sp^2 - K_2p^3 + (\partial^2p/\partial \rho^2)$$
$$\partial s/\partial \tau = a - \kappa s - sp^2 + K_2p^3 + d\,(\partial^2s/\partial \rho^2) \qquad (2.7)$$

where

$$d = D_s/D_p \qquad (2.8)$$

The dimensionless equations (2.7) may be written in the form

$$\partial p/\partial \tau = f\,(p,s) + (\partial^2p/\partial \rho^2)$$
$$\partial s/\partial \tau = g\,(p,s) + d(\partial^2s/\partial \rho^2) \qquad (2.9)$$

where f(p,s) and g(p,s) are the nonlinear reaction functions of the two dimensionless partial differential equations (2.7).

3. Complex Formation with the Activator

If the activator P (ADP) is supposed to be involved in a complexing reaction [19, 20] similar to that reported in the Turing structure experiments and modeling of the CIMA reaction [21, 22], we have the chemical equilibrium as given below.

$$P + C \leftrightarrows PC \tag{3.1}$$

where the activator P is captured partially producing the complex PC by reaction with the chemical species C. The possible nature of the complexing species is an unsolved question which is left open for future investigation. The equilibrium constant K for this complex formation reaction is given by,

$$K = pc/p.\ c \tag{3.2}$$

where pc, p, and c are the equilibrium concentrations of the complex PC, the activator P and the complexing agent C respectively. If we use the complexing agent in large excess, such that the initial concentration c_o of the complexing agent is almost equal to its concentration c at chemical equilibrium, one can define a new constant K' such that.

$$K' = K.c_o \tag{3.3}$$

Therefore, due to complexing reaction of the activator P, the new R-D equations take the forms as given below.

$$\begin{aligned} \partial p/\partial \tau' &= f(p, s) + (\partial^2 p/\partial \rho^2) \\ \partial s/\partial \tau' &= (1 + K')\,[g(p, s) + d\,(\partial^2 s/\partial \rho^2)] \end{aligned} \tag{3.4}$$

where

$$\tau = (1 + K')\,\tau' \tag{3.5}$$

4. Hopf and Turing Bifurcations

Applying the linear operator a_{ij} (k) from equation (A4) [see Appendix A], the characteristic equation for the dimensionless partial differential equations (3.4) can be written as:

$$\omega_k^2 + (k^2 d + k^2 + k^2 dK' - a_{11} - a_{22})\,\omega_k + \det[\mathbf{A}] - (a_{22} + da_{11} + dK'a_{11})k^2 + d\,(1 + K')k^4 = 0 \tag{4.1}$$

where

$$a_{11} = f_p' \quad ; \qquad a_{12} = f_s'$$
$$a_{21} = (1 + K')\, g_p' \quad ; \qquad a_{22} = (1 + K')\, g_s' \tag{4.2}$$

and from Eqs.(2.7) and (2.9), we have the following relations:

$$f_p' = -1 + 2s_o p_o - 3K_2 p_o^2$$
$$f_s' = p_o^2$$
$$g_p' = -2s_o p_o + 3K_2 p_o^2$$
$$g_s' = -\kappa - p_o^2 \tag{4.3}$$

and the stability matrix **A,** given by Eq.(4.4),

$$\mathbf{A} = \begin{bmatrix} a_{11} & a_{12} \\ a_{21} & a_{22} \end{bmatrix} \tag{4.4}$$

First, we consider this two variable system for homogeneous (k=0) mode. A Hopf bifurcation of the homogeneous system occurs when

$$\mathrm{Tr}\mathbf{A} = 0 \tag{4.5}$$

with the frequency of oscillation ω_H, and Hopf period (T_H) given by Eqs.(4.6) and (4.7) respectively.

$$\omega_H = \mathrm{mod}(\omega) = (\det \mathbf{A})^{1/2} \neq 0 \tag{4.6}$$

$$T_H = 2\pi / \omega_H = 2\pi / (\det A)^{1/2} \tag{4.7}$$

Substituting the values of a_{11} and a_{22} in Eq.(4.5) from Eqs.(4.2) and (4.3), one obtains,

$$\kappa(1 + K') + K'p_o^2 = -1 + 2s_o p_o - 3K_2 p_o^2 - p_o^2 \tag{4.8}$$

From the characteristic equation (4.1) for nonzero k mode, the condition for Hopf-wave instability is given by

$$k^2 d - a_{22} + k^2 + k^2 dK' - a_{11} \leq 0 \tag{4.9}$$

for

$$\det\mathbf{A} - (a_{22} + da_{11} + da_{11}K')k^2 + d(1+K')k^4 > 0 \tag{4.10}$$

Rearranging Eq.(4.9), we have

$$k^2_{\text{Hopf}} \leq \frac{a_{11} + a_{22}}{1+(1+K')d} \quad (4.11)$$

Since $k = 2\pi/\lambda$, where λ is the wave length of pattern, we have from Eq.(4.11), after substituting the values of a_{11} and a_{22} from equations (4.2) and (4.3), the Hopf wavelength given by the expression,

$$\lambda^2_{\text{Hopf}} \geq \frac{4\pi^2[1+d(1+K')]}{-1+2s_o p_o - 3K_2 p_o^{\,2} + (1+K')(-\kappa - p_o^{\,2})} \quad (4.12)$$

Turing bifurcation occurs when for a certain nonzero mode k of the perturbation c* (See Eq.A3), the real part of the eigenvalue ω_k of the linear operator $a_{ij}(k)$ becomes positive, so that the homogeneous steady state becomes unstable and the system undergoes a transition from homogeneous state to a perturbed state, while k=0 mode remaining stable . This may occur when the constant term of the characteristic equation (4.1) is zero. That is

$$\phi(k^2) = \det \mathbf{A} - (a_{22} + da_{11} + da_{11}K')\,k^2 + d\,(1 + K')k^4 = 0 \quad (4.13)$$

provided that

$$a_{22} + da_{11} + da_{11}K' > 0 \text{ and } a_{11} + a_{22} < 0 \quad (4.14)$$

The critical wave length (wavenumber) is determined by the degenerate root of Eq. (4.13). Thus we have

$$(a_{22} + da_{11} + dK'a_{11})^2 = 4d\,(1 + K')\det\mathbf{A} \quad (4.15)$$

Therefore, the condition for Turing instability is

$$(a_{22} + da_{11} + dK'a_{11})^2 \geq 4d\,(1+ K')\det\mathbf{A} \quad (4.16)$$

Substituting the values of a_{11}, a_{22} and det**A** from Eqs.(4.2) and the values of f_p', f_s', g_p' and g_s' from Eqs.(4.3) into Eq.(4.16), one obtains the Turing instability condition in terms of dimensionless steady state concentrations of P and S as given below.

$$[-\kappa - p_o^2 + d\,(-1 + 2s_o p_o - 3K_2 p_o^2)]^2 \geq 4d[(-1 + 2s_o p_o - 3K_2 p_o^2).(-\kappa - p_o^2) - p_o^2\,(-2s_o p_o + 3K_2 p_o^2)] \quad (4.17)$$

Combining Eqs.(4.13) and (4.15), we get the critical wave number (k_c) at the onset of Turing bifurcation as

$$k_c^2 = \left[\frac{\det A}{d(1+K')} \right]^{1/2} \tag{4.18}$$

Substituting the value of det A from Eqs.(4.2) and (4.3) and $k=2\pi/\lambda$ in the wavenumber equation (4.18), one obtains the wavelength at the onset of Turing bifurcation given by the expression

$$\lambda_c^2 = 4\pi^2 \left[\frac{d}{2s_o p_o^{\ 3} + (3\kappa K_2 - 2s_o + 1)p_o^{\ 2} - 2ks_o p_o + \kappa} \right]^{1/2} \tag{4.19}$$

The first requirement in the inequalities (4.14) is for k^2 of Turing patterns must have been positive. Substituting the values of a_{11} and a_{22} from Eqs.(4.2), the first requirement takes the form

$$g_s' + df_p' .> 0 \tag{4.20}$$

During Turing bifurcation ($k\neq0$ mode), the stability matrix **A** (k=0 mode) has to remain stable (see the second requirement in the inequalities (4.14)). We substitute the values of a_{11} and a_{22} from Eqs. (4.2) and (4.3) into this inequality to obtain

$$\kappa\,(1+K') + K'p_o^2 > -1 + 2s_op_o - 3K_2p_o^2 - p_o^2 \tag{4.21}$$

For the model glycolytic oscillation, Eq.(4.17) must be satisfied for Turing instability to occur, while the stability matrix **A** (k=0 mode) still remains stable satisfying the condition as given by Eq.(4.21).

5. Nonequilibrium Thermodynamics

Living organisms are open systems exchanging energy and matter with the surroundings. According to the second law of thermodynamics, we must have $\Delta S_{sys} + \Delta S_{surr} \geq 0$ for an organism. Let $\Delta_i S$.> 0 be the system's entropy change due to the irreversible processes in the glycolytic pathway, which occur inside the system and $\Delta_e S$ be the system's entropy change due to the exchange of energy and matter between the system and surroundings. Therefore, $\Delta S_{sys} = \Delta_i S + \Delta_e S$. The $\Delta_e S$ term must be negative to compensate for the positive $\Delta_i S$. The organism discards matter with a greater entropy content than the matter it takes in, thereby losing entropy to the environment to compensate for the entropy produced in internal irreversible processes.

For such systems the classical level of thermodynamic description is in terms of the rate of chemical entropy production (e.p.) with the associated generalized flux and thermodynamic force [25-30]. This e.p. formalism is essentially built on the validity of the local equilibrium hypothesis with respect to temperature even at nonequilibrium situations. Garcia-Colin et al. [31-33] discussed the possibility of extending this classical formalism so that the kinetic mass action law may be implemented in terms of the extended irreversible thermodynamics, for which the local equilibrium hypothesis may not apply. However, it seems acceptable that for most reactions, the thermodynamic variables change on the same time scale as the progress variable and therefore, there is no need [34] for extended thermodynamics.

A direct relation between the affinity (A) and the heat of reaction (Q) is given by [35]

$$A = -(\partial Q/\partial \xi)_{P,T} + T(\partial S/\partial \xi)_{P,T} \quad (5.1)$$

where the extent of the reaction (ξ) is related to its velocity (J) by the relation

$$J = d\xi/dt \quad (5.2)$$

and the subscripts P and T indicate the constancy of pressure and temperature, respectively. It is sometimes possible to neglect the entropy (S) variation term in Eq. (5.1) because of its minor contribution. In such cases, the entropy production due to a chemical change becomes simply proportional to the heat of reaction as given below.

$$d_i S/dt = AJ/T = -(1/T)(\partial Q/\partial t)_{P,T} \quad (5.3)$$

In this approximation, the e.p. of a living organism can be measured by its metabolism as recorded by calorimetry.

A linear thermodynamic analysis of this model system close to thermodynamic equilibrium at constant temperature has been undertaken [15] by an e.p. technique of Prigogine[27] to identify that only the autocatalytic step (2.2) can destabilize the steady states.

For the thermodynamic description, the rate equations [35] for the steps (2.1) to (2.3) of the model (II) using the scaling defined in Eqs. (2.5) are given by,

$$\begin{aligned} J_1 &= k_3N(a - \kappa s) \\ J_2 &= k_3N(sp^2 - K_2p^3) \\ J_3 &= k_3N(p - b) \end{aligned} \quad (5.4)$$

where J_i represents the reaction velocity of the ith step. Since the concentrations of ATP and ADP are very low (~ of the order of less than 1 mM), the ideal solution approximation for the mixture of reactant and product molecules is applicable. Also, using the scaling defined in Eqs. (2.5), one obtains the affinities[35] of the individual steps (A_i) and the overall reaction (A) given by the relations,

$$A_1 = RT \ln (a/\kappa s)$$
$$A_2 = RT \ln (s/K_2p)$$
$$A_3 = RT \ln (p/b)$$
$$A = RT \ln (a/\kappa K_2b) \qquad (5.5)$$

The time-independent steady states are described by the following two equations:

$$\kappa s_o = (a + b) - p_o$$
$$(1 + \kappa K_2)p_o^3 - (a + b)\, p_o^2 + \kappa(p_o - b) = 0 \qquad (5.6)$$

Therefore, one obtains the steady state values (s_o, p_o) numerically from Eqs.(5.6) for the known values of the parameters κ, K_2, a, and b. The e.p. per unit volume (d_iS/dt) due to nonequilibrium steady states is given by

$$d_iS/dt = (1/T) \sum_i J_i A_i \geq 0 \qquad (5.7)$$

where J_i and A_i are, respectively, the velocity and affinity of the ith reaction. Substituting the values of J_i and A_i from Eqs.(5.4) and (5.5) to scale time t to τ, one obtains

$$d_i S/d\tau = NR\,[\,(a - \kappa s) \ln (a/\kappa s) + (sp^2 - K_2p^3) \ln (s/K_2p) + (p - b) \ln (p/b)\,] \qquad (5.8)$$

Once the values of a, b, κ, K_2, and the steady state values, s_o, and p_o are known, one can calculate $d_iS/d\tau$ from Eq.(5.8) for nonequilibrium states including the state of thermodynamic equilibrium represented by Eqs.(5.9) as given below. The nonlinear oscillatory region has been explored in the past by Richter et al. [14] with the following interesting results – the e.p. in the oscillatory state may be greater or less than that in the corresponding unstable steady state - there is no consistent relationship between the e.p. in the oscillatory state and that in the corresponding unstable steady state.

The following relations [14, 15] hold good at the state of thermodynamic equilibrium of the reaction scheme (II); subscript 'e' stands for the equilibrium state.

$$s_e = a/\kappa \;;\; p_e = s_e/K_2 = a/\kappa K_2 \;;\; p_e = b = a/\kappa K_2 \qquad (5.9)$$

It is worthwhile to mention that the time-independent steady state coincides with the state of thermodynamic equilibrium if the boundary conditions are compatible with the equilibrium condition i.e., if the flux of the open system is made zero (closed system).

6. Results and Discussion

Hopf bifurcation of the homogeneous (k=0) and inhomogeneous ($k \neq 0$) model system is strongly affected by complex formation with the activator P — the Hopf boundary would be shifted to a lower value of κ, if K' has values >0. This is presented in figure 1 with two more values of K'>0 (0.5 and 1.5 respectively). Figure 2 demonstrates how an unstable steady state 'O' bifurcates to a stable limit cycle in s-p phase plane due to Hopf bifurcation. The negative Floquet exponents calculated[16,36-39] along the Hopf boundary shown in figure 1 (K'=0) ensure that the bifurcations are supercritical generating only stable limit cycles at least in the parameter range we have studied.

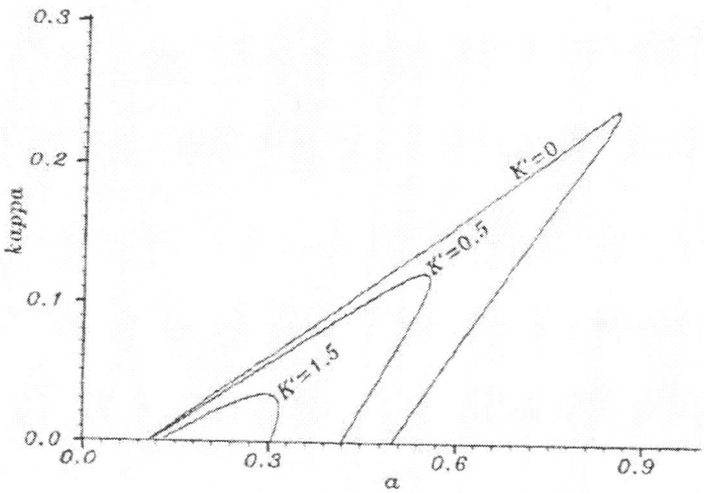

Figure1. Phase diagram in the (a - κ) phase plane of the homogeneous (k=0) reversible Sel'kov model for K'= 0, 0.5, and 1.5 respectively : b, 0.09 ; K_2, 1 ; a and b are the scaled concentrations of source A and sink B, respectively.

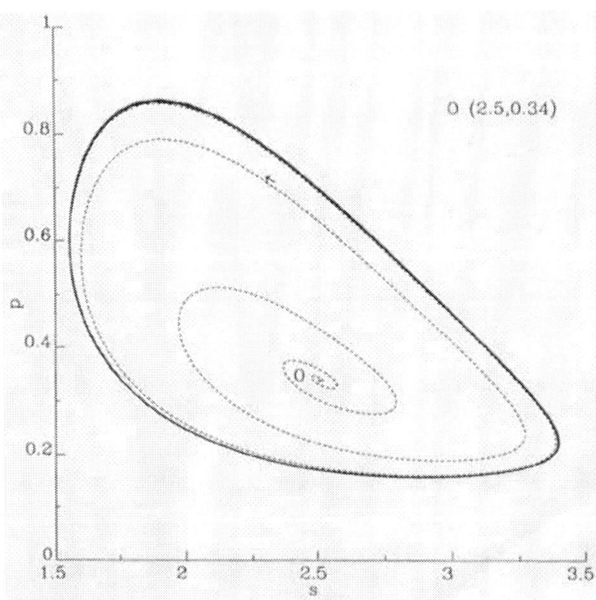

Figure2. A plot of limit cycle oscillations in the s-p plane of the homogeneous reversible Selkov model: oscillation period T, 22.25; a, 0.5 ; b, 0.09 ; κ, 0.1, K_2, 1 ; 'O' is an unstable steady state, which bifurcates to a limit cycle.

In the case of Hopf-wave bifurcation ($k \neq 0$ mode), the wavenumber (wavelength) depends strongly on the degree (K') of the complexing reaction of the activator P. This is evident from Eq. (4.11) and also in figures 3a and 3b in reference 18, suggesting that due to the complexing reaction of the activator (K', 0.5), there is an approximately two-fold decrease (increase) of Hopf wavenumber (wavelength) values. When translated to real lengths, our model calculations predict inhomogeneous Hopf-waves on the order of a few millimetres in length, which compares well with the work of Müller et al[40], who have reported polygonal network patterns (size approximately 0.5 – 1 mm) due tp Marangoni effect in a R-D experiment of glycolysis in yeast extracts. The values predicted from another R-D model of glycolysis [41,42] is about 3 mm in length.

The wavenumber k=0 (wavelength, infinitely large) should make the reaction medium perfectly homogeneous, which can be attained in a real experiment only at a very high stirring rate, where molecular diffusion has no chance to act. A decrease in stirring initiates the role of molecular diffusion as an important transport process, introducing inhomogeneity from imperfect mixing [43,44]. This inhomogeneity manifests itself in the form of nonzero wavenumber modes in an unstirred R-D medium, where the wavenumber may attain the appropriate value including the maximum possible value (wavelength shortest) as decided solely by the bifurcation parameters. We have made a numerical estimate[18] of the domain of different nonzero wavenumber modes in the parameter space for low, moderate, and high values of d in the absence (K', 0) and presence (K'>0) of the complexing reaction of the activator P. Figures 3 and 4 of reference 18 indicate that the areas of phase diagrams in the parameter space and the corresponding maximum values of the possible wavenumber k (shortest wavelength) are reduced drastically at high values of d as well as in the presence of the complexing reaction of the activator P. This result is in agreement with that obtained in figures 3a and 4b from k_H versus a plots in the absence and presence of the complexing reaction with the activator P, respectively [18].

Figures 4a, and 4b demonstrate [45] that, while the parameter κ is being decreased gradually, Turing region preceeds the Hopf bifurcation line (k=0) in the (a - κ) phase plane both in the absence and presence of complexing reaction with the activator P. It is interesting [45] to note that, for a very low value of d = 5, the Hopf-domain, in the parameter space, decreases with the increase of K' (degree of complex formation with the activator P), but the tiny Turing region remains practically unchanged in size. Since a very low value of d, for the parameters chosen, does not help the Turing domain expand for K'>0, a gap between these two domains is noticeable for low values of d [45]. For large values of $d \geq 10$, however, Turing domain expands in the parameter space at the cost of Hopf domain for K'>0, and apparently there exists no gap of seperation between these two domains as presented in figures 4a and 4b..

But the complex formation reaction (K'>0) with the activator species P neither changes the wavelength of Turing patterns (see equation 4.19) nor does affect the basic relation to be maintained for Turing instability to occur (see equation 4.17). Turing wavenumber k_c as a function of a is presented in figures 5a and 5b for different values of d [45] in the absence and presence of the complexing reaction of the activator, respectively, to verify that k_c , indeed does not depend on K'. It is interesting to note from these two plots that K'=1 does help to obtain Turing wavenumbers (k_c) at a very low value of d=5 which, otherwise could not produce any Turing structures in the absence of any complexing reaction with the activator (K', 0). This result is in agreement with the observations of De Kepper et al.[46] on the CIMA reaction with varying concentrations of starch. Also, the inequality (4.20), which ensures the positivity of the value of k^2 for Turing patterns to occur, has no term containing K'. The only link of K' with the Turing structure formation is obviously the inequality (4.21), which must be satisfied simultaneously with the inequality (4.20) and the basic relation (4.17). By increasing K' in the inequality (4.21) to an appropriate value, it is possible to arrest the arrival of Hopf bifurcation in this model system for k=0 mode (see figure 1), and the Hopf domain which disappears by this technique, may be used [45] to generate Turing patterns (see figures 4 and 5) by inducing inhomogeneous perturbations of nonzero k mode.

The computed Turing wavenumbers in figures 5a and 5b, when translated to real unit of length, give the Turing wavelength values in the range 0.21 – 0.15 mm, which is left open for comparison with the values from the real experiments of glycolytic Turing patterns in the future

Different higher values of d can be realized in experiments by adopting a regulatory mechanism based on immobilization of the activator P in a real glycolytic system. Therefore, the experimental wavelengths of the Turing patterns and Hopf-waves including target patterns, spiral waves etc. at different values of d and bifurcation parameters may be used for comparison, when available, to test the findings of the numerical predictions reported here.

Figure 6 shows the affinity as a function of the parameter a for the individual steps as well as the overall reaction for the appropriate values of b, κ, and K_2, which yield the most convenient results (see Eqn. 5.5) . The affinity of steps (1) and (3) of this model (II) increases slowly with the increase of a; but in the case of the autocatalytic step (2), the curve first shows a maximum [47] with the increase of a, and decreases thereafter with further increase of a. The affinity curve for the overall reaction shows an exponential increase with the increase of a, indicating that a is a measure of the distance from the state of thermodynamic equilibrium (see Eqs. 5.9). Figure 7 is the variation of velocities (J_i) of the three individual steps of the model (II) with their corresponding affinities (A_i). The affinities of steps (1) and (3) increase exponentially with the increase of the velocities of the corresponding steps, whereas for the autocatalytic step (2), the reaction velocity remains very very low [47] until the corresponding affinity overcomes a barrier of ~2.95RT, beyond which the velocity of step(2) increases exponentially accompanied with a decrease of its affinity. Interestingly, the J_2 versus A_2 plot in figure 7 displays a beautiful allosteric regulatory mechanism of the autocatalytic step (2). The total velocity J increases exponentially with the total affinity A, has been reported in the past (see figure 3 of reference 14). Figure 8 presents the e.p. as a function of the overall affinity A for the most convenient values of κ, K_2, and b. This gives a numerical estimate of the e.p. in this model (II) at different distances from the state of thermodynamic equilibrium (figure 4 of reference 14 has presented the calculation of e.p.of steps (1), (2), and (3) of this model (II) separately).

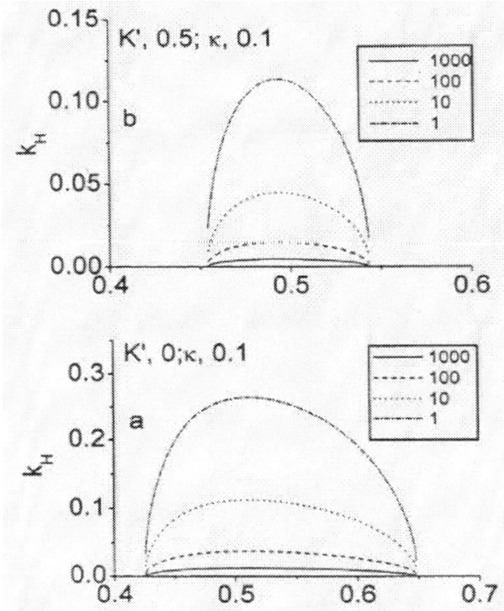

Figure 3. Hopf wavenumber (k_H) as a function of a for four different values of d ; κ, 0.1 ; other parameters same as Figure 1 : (a) K', 0 ; (b) K', 0.5.

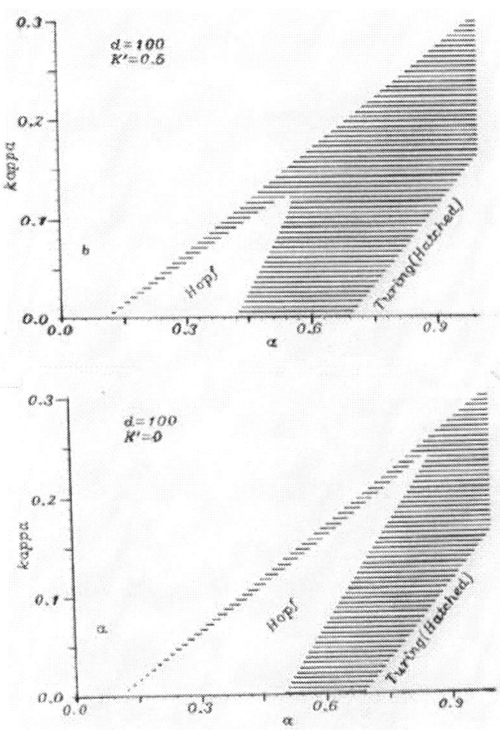

Figure 4. Phase diagram in the (a - κ) phase plane of the reversible Sel'kov model for d=100 in absence (K', 0) and presence (K', 0.5) of complexing reaction of the activator P; white and hatched areas are Hopf and Turing regions respectively; other parameters, same as Fig 1: (a) K′, 0; (b) K′, 0.5.

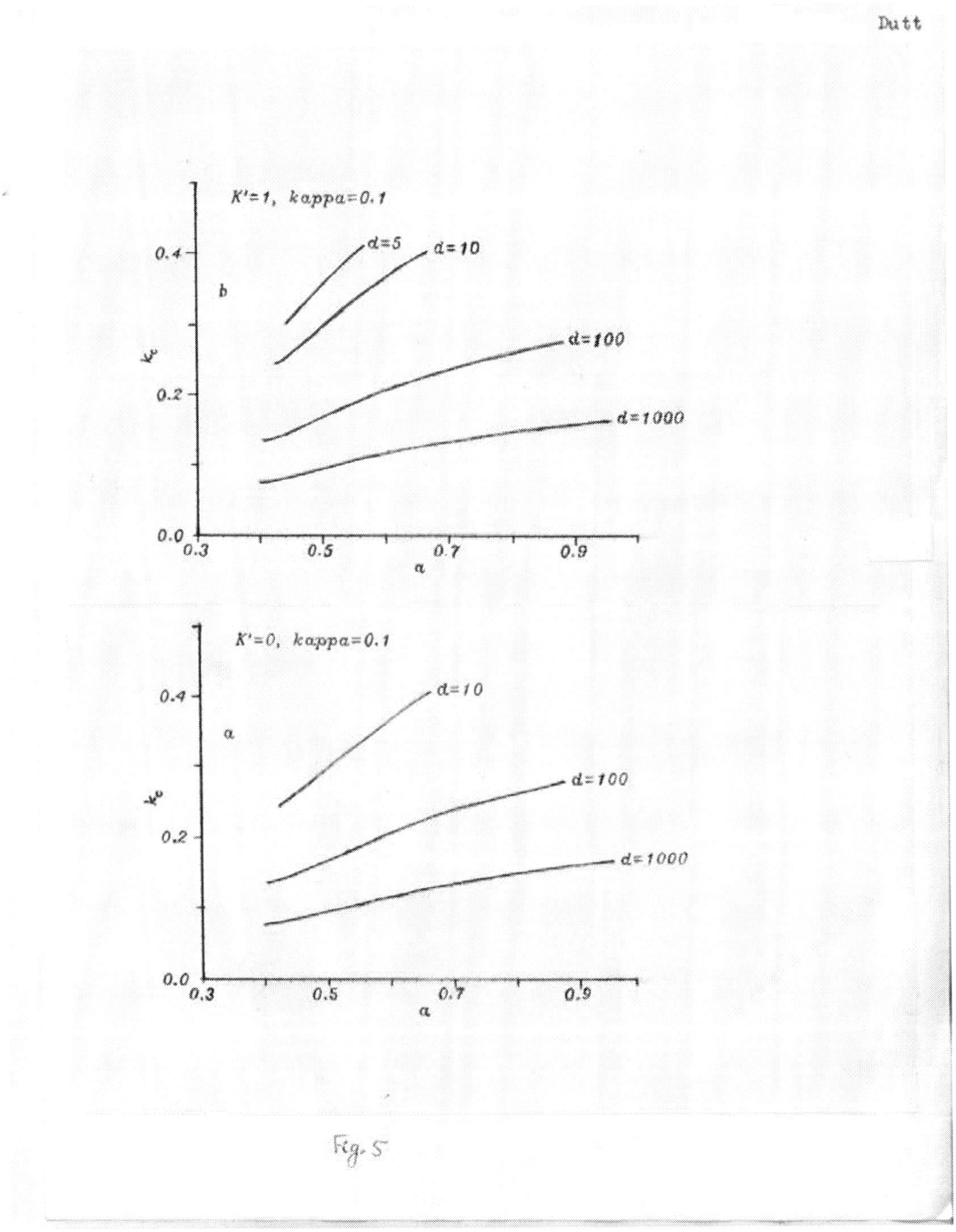

Figure 5. Turing wavenumber (k_c) as a function of a for four different values of d ; κ, 0.1 ; other parameters, same as Figure 1 : (a) K′, 0 ; (b) K′, 1.

An application of the calculation of e.p. in glycolytic pathway is – the less the dissipation (T times e.p. for this model isothermal process), the less is the Gibbs free energy change

(ΔG at constant T and P), and the more is the energy transduction from reactants to products to ensure higher efficiency of the biochemical engines. Figures 6 and 8 demonstrate that the low values of a (the scaled concentration of the input substrates such as glucose and fructose) in the range $0 < a < 0.3$, which correspond to the overall affinity (A) in the range $0 < A/RT < 3.5$, produce very low e.p. (dissipation). This low e.p. region may be successfully utilized to ensure higher energy transduction in a stationary glycolytic pathway.

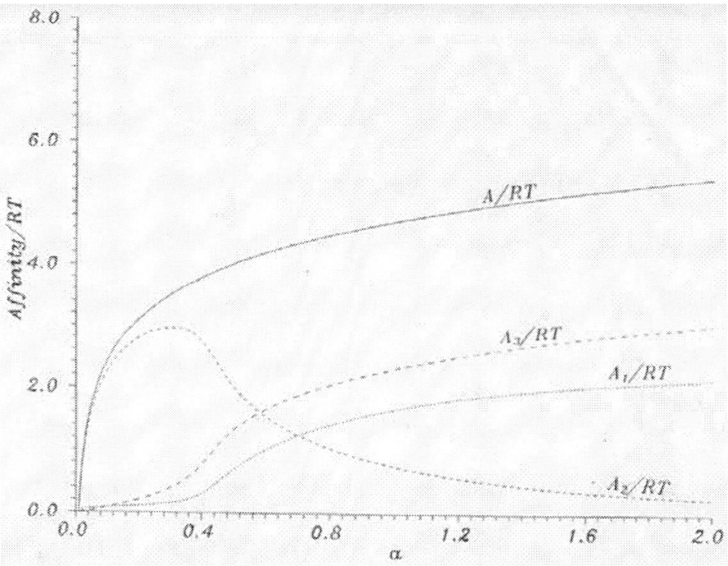

Figure 6. The affinity as a function of a in the reversible Sel'kov model : A_i, the affinity of the ith step of model(II) (i = 1, 2, and 3 respectively) and A, the overall affinity ; κ, 0.1 ; K_2, 1 ; b, 0.09.

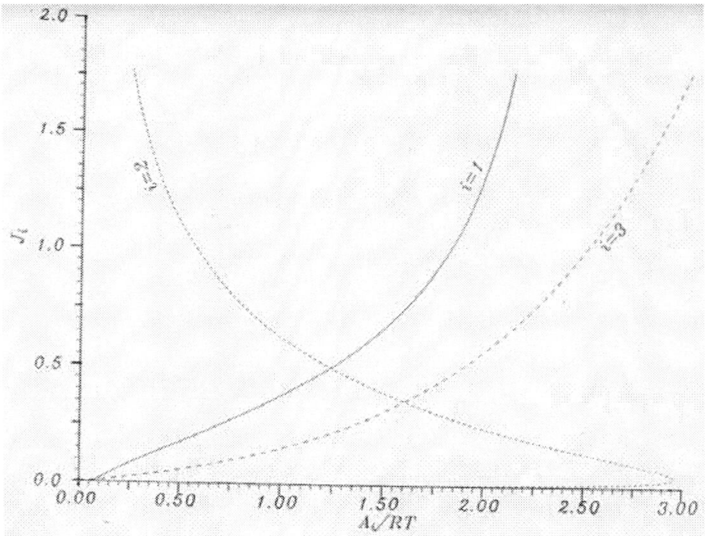

Figure 7. The velocity (J_i) as a function of the affinity (A_i) of the three individual steps of the model(II); the parameters are the same as figure 6.

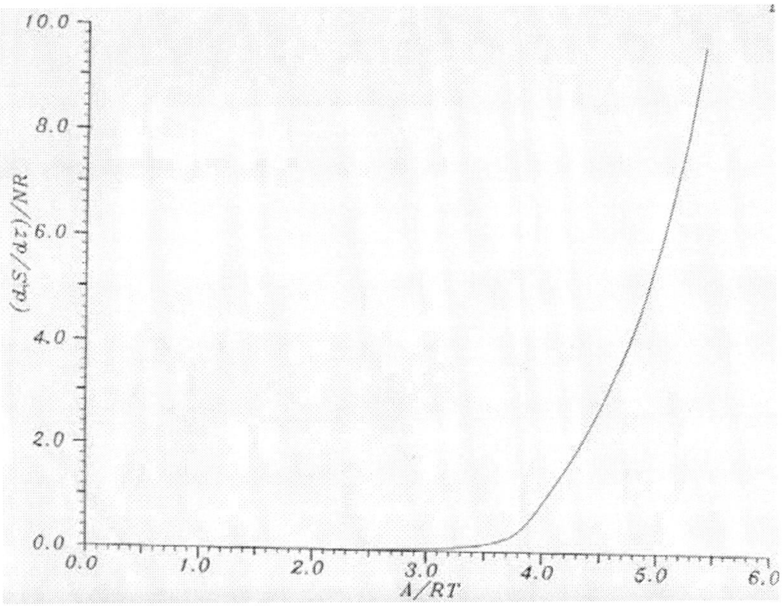

Figure 8. The entropy production as a function of the overall affinity (A) in the model (II); the parameters are the same as figure 6.

This paper has presented a review of the nonlinear dynamics and nonequilibrium thermodynamics in the reversible Sel'kov model, a mathematical model of glycolytic oscillations. The discussion on nonequilibrium thermodynamics is based on chemical reactions from the steps (2.1 - 2.3) of this model system. The discussion on nonlinear dynamics has included stationary Turing patterns and Hopf-waves separately – we have not considered any kind of interaction [48-50] between them. Interaction and competition between these two symmetry breaking modes is expected to generate a large variety of spatiotemporal patterns[51] including modulated Turing structures, modulated stationary waves and a combination of Turing structures and spiral waves in this model system (II)[13,14] of glycolytic oscillations.

7. Acknowledgment

I acknowledge EPSRC for financial support in part.

Appendix A

Linear Stability Analysis: Let us describe a reaction-diffusion (R-D) system in two variables by

$$\partial u/\partial t = f(u,v) + D_u (\partial^2 u/\partial \mathbf{r}^2) \quad \text{(A1)}$$
$$\partial v/\partial t = g(u,v) + D_v (\partial^2 v/\partial \mathbf{r}^2) \quad \text{(A2)}$$

where u and v are the chemical concentrations of the species U and V participating in the reaction described by the nonlinear functions f(u,v) and g(u,v) and D's are the diffusion coefficients of the species U and V. We assume that the Eqs.(A1) and (A2) have a homogeneous steady state solution $f(u_o,v_o) = 0$ and $g(u_o,v_o)=0$. We consider the evolution of a small perturbation c^* around the steady state concentration c_o and separate it in Fourier space

$$c^* = \delta c \sum_k a_k \exp(\omega_k t + i\mathbf{k}.\mathbf{r}) \quad \text{(A3)}$$

where ω_k is the growth rate of the mode with a wavevector **k**. We substitute Eq. (A3) into Eqs.(A1) and (A2) and expand it by Taylor series for two variables around the homogeneous steady state (u_o,v_o). Retaining only the linear terms for the homogeneous steady state condition $f(u_o,v_o)=0$ and $g(u_o,v_o)=0$, we obtain an eigenvalue equation [23,24] for ω_k for the linear operator

$$a_{ij}(k) = a_{ij} - D_i k^2 \delta ij \quad \text{(A4)}$$

The derivation of Equation (A4) is given in the Appendix B. Here we have assumed that without the diffusion term, all the eigen values of the stability matrix **A** have negative real parts and the steady state is stable. Diffusion-driven instability is the problem of finding the condition for which the matrix **A** (k) = **A** – k^2**D** has an eigenvalue with positive real part for nonzero D's.

Appendix B

According to Eq.(A3), let

$$u = u_o + \delta u \sum_k a_k \exp(\omega_k t + i\mathbf{k}.\mathbf{r}) \quad \text{(A3a)}$$

$$v = v_o + \delta v \sum_k a_k \exp(\omega_k t + i\mathbf{k}.\mathbf{r}) \quad \text{(A3b)}$$

Substituting the values of $\partial u/\partial t$, $\partial^2 u/\partial \mathbf{r}^2$, $\partial v/\partial t$, $\partial^2 v/\partial \mathbf{r}^2$ in the R-D equations (A1) and (A2), one obtains for the first partial differential equation (A1),

$$\omega_k \delta u \sum_k a_k \exp(\omega_k t + i\mathbf{k}.\mathbf{r}) = f\{[u_o + \delta u \sum_k a_k \exp(\omega_k t + i\mathbf{k}.\mathbf{r})],$$

$$[vo + \delta v \sum_k a_k \exp(\omega_k t + i\mathbf{k}.\mathbf{r})]\} - k^2 D_u \delta u \sum_k a_k \exp(\omega_k t + i\mathbf{k}.\mathbf{r}) \quad \text{(A3c)}$$

Expanding the first term of the right hand side by Taylor series for two variables and retaining only the linear terms, one obtains for homogeneous steady state condition,

$$\omega_k \delta u = f_u'\delta u + f_v'\delta v - k^2 D_u \delta u \quad \text{(A3d)}$$

Similarly for the second R-D equation (A2), one obtains

$$\omega_k \delta v = g_u'\cdot\delta u + g_v'\delta v - k^2 D_v \delta v \quad \text{(A3e)}$$

Combining Eqs.(A3d) and (A3e), the characteristic equation is given by

$$\text{Det}\begin{bmatrix} f_u' - k^2 D_u - \omega_k & f_v' \\ g_u' & g_v' - k^2 D_v - \omega_k \end{bmatrix} = 0 \quad \text{(A3f)}$$

Therefore, the linear stability matrix in presence of diffusion takes the form

$$\begin{bmatrix} f_u' - k^2 D_u & f_v' \\ g_u' & g_v' - k^2 D_v \end{bmatrix} = \begin{bmatrix} f_u' & f_v' \\ g_u' & g_v' \end{bmatrix} - k^2 \begin{bmatrix} D_u & 0 \\ 0 & D_v \end{bmatrix} \quad \text{(A3g)}$$

Therefore,

$$a_{ij}(k) = a_{ij} - D_i k^2 \delta ij \quad \text{(A4)}$$

References

[1] Olsen, L.F. ; Degn, H. *Nature,* 1977, 267, 177.

[2] Olsen, L.F. ; Degn, H. B*iochim. Biophys. Acta,* 1978, 523, 1321.

[3] Degn, H. ; Olsen, L.F. ; Perram, J.W. *Ann. N.Y. Acad. Sci.* 1979, 316, 622.

[4] Olsen, L.F. *Phys. Letts.A,* 1983, 94, 454.

[5] Yamazaki, I. ; Yokota, K. ; Nakajima, R. *Biochem. Biophys. Res. Commun.* 1965, 21, 582.

[6] Nakamura, S. ; Yokota, K. ; Yamazaki, I. *Nature*, 1969, 222, 794.

[7] (a) Larter, R. ; Bush, C.L. ; Ronis, T.R. *J.Chem.Phys.* 1987,.87, 5765 , (b) Scheeline, A. ; Olsen, D.L. ; Williksen, E.P. ; Horras, G.A. ; Klein, M.L. ; Larter, R. *Chem.Rev.* 1997, 97, 739.

[8] Hess, B. ; Boiteaux, A. *Ann.Rev.Biochem.* 1971, 40, 237.

[9] Hess, B. ; Boiteaux, A. ; Kruger, *J. Adv. Enzyme Regul.* 1968, 7, 147.

[10] Hess, B. ; Plesser, T. *Ann.N.Y.Acad.Sci.,* 1979, 316, 203.

[11] Markus, M. ; Hess, B. *PNAS (USA),* 1984, 81, 4394.

[12] Termonia, Y. ; Ross, J. *PNAS, USA,* 1981, 78, 2952.

[13] Selkov, E.E. ; *Eur.J.Biochem.* 1968, 4, 79.

[14] Richter, P. ; Regmus, P. ; Ross, J. *Prog.Theor.Phys.* 1981, 66, 385.

[15] Dutt, A.K. *J.Chem.Phys.* 1990, 92, 3058.
[16] Dutt, A.K. *Chem.Phys.Letts.* 1993, 208, 139.
[17] Dutt, A.K. *Chem.Phys.Letts.* 2002, 357, 341.
[18] Dutt, A.K. *J.Phys.Chem.B,* 2005, 109,17679.
[19] Turing, A. *Philos. Trans.. R. Soc.* (London) B, 1952, 237, 37.
[20] (a) Epstein, I. ; Lengyel, I. Physica D, 1995, 84, 1 ; (b) Lengyel, I. ; Epstein, I. *PNAS(USA),* 1992, 89, 3977.
[21] De Kepper, P.; Epstein, I. ; Orban, M. ; Kustin, K. *J.Phys.Chem.* 1982, 86, 170.
[22] Castets, V. ; Dulos, E. ; Boissonade, J. ; De Kepper, P. Phys.Rev.Lett. 1990, 64, 2953.
[23] Murray, J. *Mathematical Biology.* (Springer, Berlin,1989)
[24] Qian, H. ; Murray, J. *Appl.Math.Lett.* 2001, 14, 405.
[25] De Groot, S.R. ; Mazur, P. Nonequilibrium Thermodynamics. (North Holland, Amsterdam,1969).
[26] Nicolis, G. ; Prigogine, I. Self-Organization in Nonequilibrium Systems (*Wiley-Interscience,* New York, 1977).
[27] Glansdorff, P. ; Prigogine, I. Thermodynamic Theory of Structure Stability and Fluctuations (*Wiley-Interscience,* New York, 1971).
[28] Keizer, J. *Acc. Chem. Res.* 1979, 12, 243.
[29] Keizer J. Statistical Thermodynamics of Nonequilibrium Processes (*Springer,* New York, 1987).
[30] Casas-Vasquez, J. ; Jou, D. ; Lebon, G. Recent Developments in Nonequilibrium Thermodynamics (*Springer,* New York, 1984).
[31] Garcia-Colin, L. ; De La Selva, S. J. *Nonequilib. Thermodyn.* 1983, 8, 277.
[32] Garcia-Colin, L. ; De La Selva, S. ; Pine, E. *J.Phys.Chem.* 1986, 90, 953.
[33] De La Selva, M. ; Garcia-Colin, L. *J. Chem.Phys.*1986, 85, 2140.
[34] Ross, J. ; Garcia-Colin, L. *J. Phys. Chem.* 1989, 93, 2091.
[35] Prigogine, I. Thermodynamics of Irreversible Processes (*Wiley-Interscience,* New York, 1967).
[36] Poore, A.B. *Arch.Rat.Mech.Anal.* 1976, 60, 371.
[37] Ashkenazi, M. ; Othmer, H.G. *J.Math.Biol.* 1978, 5, 305.
[38] Othmer, H.G. ; Aldridge, J.A. *J.Math.Biol.* 1978, 5, 169.
[39] Guckenheimer, J. ; Holmes, P. Nonlinear Oscillations, Dynamical Systems and Bifurcations of Vector Fields (*Springer,* Berlin, 1983).
[40] Müller, S. ; Plesser, T. ; Boiteaux, A. ; Hess, B. Z.Naturforsch.C; *J.Biosci.* 1985, 40, 588.
[41] Goldbeter, A. ; Lefever, R. *Biophys.J.* 1972, 12, 1302.
[42] Goldbeter, A. *PNAS,USA,* 1973, 70, 3255.
[43] Epstein, I. *Nature,* 1990, 346, 16.
[44] Dutt, A.K. ; Datta, A. *J.Phys.Chem.A,* 1998, 102, 7981.
[45] Dutt, A.K.(Submitted for publication).
[46] Agladze, K. ; Dulos, E. ; De Kepper, P. *J.Phys.Chem.* 1992, 96, 2400.
[47] Dutt, A.K. *J. Phys. Chem.A.* 2002, 106, 2401.
[48] Vanag, V. ; Epstein, I. *Science,* 2001, 294, 835.
[49] Vanag, V. ; Epstein, I. *Phys. Rev. Letts.* 2001, 87, 22830.

[50] Yang, L. ; Dolnik, M. ; Zhabotinsky, A. ; Epstein, I.. *J.Chem.Phys.* 2002, 117, 7259.
[51] J.-P. Voroney, J.-P. ; Lawniczak, A.T. ; Kapral, R. *Physica D,* 1996, 99, 303

In: Glycolysis: Regulation, Processes and Diseases
Editor: Paul N. Lithaw
ISBN: 978-1-60741-103-1
© 2009 Nova Science Publishers, Inc.

Chapter IV

The Role of Skeletal Muscle Glycolysis in Whole Body Metabolic Regulation and Type 2 Diabetes

Jørgen Jensen*

Department of Physiology, National Institute of Occupational Health, Oslo, Norway, and Department of Physical Performance, Norwegian School of Sports Sciences, Oslo, Norway

Abstract

For most human at least 50 % of the dietary energy comes from carbohydrates. Skeletal muscles make up 30-40 % of the body weight and the major part of the carbohydrate is stored as muscle glycogen (≈ 80 %). After a carbohydrate meal ≈ 35 % of the carbohydrates are stored as muscle glycogen whereas 20 % ends up as liver glycogen. A major part of ingested carbohydrates, therefore, passes through glycolysis in skeletal muscles. Glycolysis in skeletal muscles is activated during insulin-mediated glucose disposal and more in muscles with high glycogen content. Skeletal muscles cannot release glucose molecules because glucose 6-phosphatase is lacking. However, muscle glycogen can be metabolised via glycolysis and released as lactate; skeletal muscles are the major contributor of blood lactate appearance. The released lactate is the major substrate for gluconeogenesis or oxidation in some tissues. Adrenaline-mediated glycogen phosphorylase activation initiates glycolysis in resting skeletal muscles. Skeletal muscle glycolytic rate is highest during high intensity exercise when muscles convert chemical energy to movement. During exercise, glycolytic rate in skeletal muscles can increase more than 100-fold, and substantial amount of glycogen is rapidly broken down in muscles. In the present paper, regulation of glycolysis in skeletal muscles by insulin, adrenaline and exercise is discussed. Furthermore, the physiological

* For correspondence: Jørgen Jensen, Department of Physiology, National Institute of Occupational Health, P. O. Box 8149 Dep., N-0033, Oslo, Norway. Fax: (+47) 23 19 52 04. Phone (+47) 23195243. E-mail: jorgen.jensen@stami.no

role of skeletal muscles glycolysis for whole body metabolic regulation in normal and type 2 diabetes is addressed.

Introduction

Skeletal muscle is the specialized tissue for movement where chemical energy is transformed into mechanical work. Skeletal muscles have, however, also a pivotal role for regulation of blood glucose. Skeletal muscles are our largest tissue making up ≈ 40 % of the body weight and 70-90 % of insulin-stimulated glucose disposal is incorporated into muscle glycogen (Shulman *et al.*, 1990). Skeletal muscles also store the majority of the body's carbohydrates. The glycogen content in skeletal muscles reaches ≈ 100 mmol/kg (≈ 16 g/kg) and skeletal muscles can therefore store about ≈ 500 g of carbohydrate in a young man of about 70 kg. The liver contains about 100 g glycogen (Taylor *et al.*, 1996).

During muscle contractions, chemical energy is transformed into mechanical work by myosin and actin. The utilization of ATP can be extremely high (Connett & Sahlin, 1996) but efficient regeneration prevents any significant decline in ATP. ATP is rapidly regenerated from CrP, but if larger amount of energy is required, additional energy is supplied via glycolysis and oxidative phosphorylation. The carbohydrate used during muscle contraction is mainly glycogen. Skeletal muscles can also take up glucose from the blood, but at a much lower rate (Coyle, 1995). Glycogen and glucose are initially metabolized through glycolysis to provide energy for ATP re-synthesis. If sufficient oxygen is present, the pyruvate produced will enter the mitochondria for complete oxidation to CO_2 and H_2O. However, at high intensities or when insufficient oxygen is present, pyruvate becomes converted to lactate.

Carbohydrates are metabolized via glycolysis for release of energy. Furthermore, ingestion of carbohydrates increases metabolism of glucose (Flatt, 1995). After a carbohydrate meal, Woerle et al. reported that 43 % was oxidized during the next 6 h (Woerle *et al.*, 2003). However, a larger part of the ingested glucose was metabolized via glycolysis in the postprandial period (66 %) but was reconverted to glucose via gluconeogenesis and incorporated into glycogen (Woerle *et al.*, 2003). Insulin increases glycolysis in skeletal muscles (Dimitriadis *et al.*, 1992;Jensen *et al.*, 2006). Adrenaline also increases glycolysis in skeletal muscles (Challiss *et al.*, 1986;Jensen & Dahl, 1995). Adrenaline stimulates glycogen breakdown and causes accumulation of lactate in skeletal muscles (Aslesen & Jensen, 1998;Jensen & Dahl, 1995;Jensen *et al.*, 1997). Skeletal muscles release the majority of the lactate into the blood, where it becomes available as energy source for other tissues (e.g. heart) or precursors for gluconeogenesis in liver and renals. In fact, the title of the papers where Cori and Cori first described gluconeogenesis was "The mechanism of epinephrine action". Cori and Cori showed that injection of adrenaline to fasted rats reduced glycogen content in muscles whereas glycogen accumulated in the liver (Cori & Cori, 1928).

Glycolysis in skeletal muscles is exceptional because the flux can increase by more than 100-fold during exercise (Connett & Sahlin, 1996). Some of the glycolytic enzymes are regulated by phosphorylation, but allosteric regulation of 6-phosphofructokinase-1 (PFK-1) is the key regulator of glycolytic flux. Glycolysis in skeletal muscles is regulated by energy requirement and substrate supply. Substrates for glycolysis come from degradation of glycogen and uptake of glucose. Glycogen phosphorylase and glucose transport are therefore important regulators of glycolytic flux.

Glycolysis is the initial step in the conversion of glucose to lipid, and glycolysis can under this condition be viewed at an anabolic reaction; the first step in lipid synthesis from glucose. The body's carbohydrate store (glycogen) is limited and excess carbohydrates have to be metabolized via glycolysis for regulation of blood glucose; furthermore, pyruvate must be oxidized or converted to lipids and excess of carbohydrate reduces fat oxidation (Flatt, 1995). However, glycolysis in skeletal muscle has also been suggested to be regulate by excess availability fat in the so-called "glucose fatty acid cycle" (Randle's cycle). In the present chapter, I will discuss regulation of glycolysis in skeletal muscles by insulin, adrenaline and contraction in relation to whole body glucose metabolism and type 2 diabetes.

Glycolysis

Glycolysis is normally considered as the chain of reactions where one glucose molecule is broken down to two pyruvate molecules via 10 enzymatic reactions. Muscles cell, like most other cells, contains only a scant amount of free glucose. On the other hand, muscles contain a substantial amount of glycogen. Therefore, physiological glycolysis starts by transport of glucose into the cells or by degradation of glycogen to glucose 1-phosphate (Figure 1).

Pyruvate, the end product of glycolysis, can be metabolised in many ways depending on cell types and metabolic status. Some cell types do not have mitochondria. In most cell types the majority of pyruvate will enter the mitochondria for oxidation. During anaerobic conditions (e.g. heavy exercise or during hypoxia), pyruvate becomes converted to lactate. In glycolysis NAD^+ is reduced to NADH and regeneration of NAD^+ is mandatory. When oxygen is sufficient, NADH is transferred into the mitochondria and enters the electron transfer chain for oxidation. With insufficient oxygen, NADH can be oxidised by lactate dehydrogenase coupled to conversion of pyruvate to lactate. Pyruvate can also be converted to oxaloacetate and contribute to anaplerosis or gluconeogenesis.

The flux through a metabolic pathway is determined by activities of rate-limiting enzymes. As mentioned above, the substrate for glycolysis can either be blood glucose or glycogen. Therefore, glycolysis can be considered to have two substrates (glycogen and glucose) and regulation of the initial steps differ under these conditions. For glycolytic breakdown of blood glucose, glucose must be transported into the cells before it becomes phosphorylated and metabolised to pyruvate. In glycolysis from glucose, the following steps can be rate-limiting: 1) glucose transport, 2) hexokinase, 3) 6- phosphofructokinase, and 4) pyruvate kinase. For glycolysis from glycogen, glycogen phosphorylase accounts for the breakdown of glycogen and supplies glucose 1-phosphate. Therefore, only three rate-limiting steps are operating: 1) glycogen phosphorylase, 2) 6- phosphofructokinase, and 3) pyruvate kinase. See Figure 1.

6-Phosphofructokinase (PFK-1) is considered as the most important regulator of glycolysis. Three genes code muscle (PFKM), liver (PFKL) and platelet (PFKP) isoforms of PFK-1. PFK-1 has a complex regulation, which includes allosteric regulation (by many metabolites), pH-mediated regulation, phosphorylation, complex formation and reversible binding to the cytoskeleton (Connett & Sahlin, 1996;Kemp & Foe, 1983). PKF-1 is allosterically activated by fructose 2,6-bisphosphate, fructose 6-phosphate, glucose 1,6-bisphosphate, AMP, P_i (Kemp & Foe, 1983;Newsholme & Leech, 1983). Allosteric inhibitors of PFK-1 are: ATP, citrate and CrP (Kemp & Foe, 1983;Newsholme & Leech, 1983). The active form of PFK-1 is a tetrameric comples, but PFK-1 can dissociates to dimers and monomer (Kemp & Foe, 1983); allosteric regulators influence complex formation. PFK-1 activity is also regulated by reversal interaction with structural proteins; most likely F-actin (Luther & Lee, 1986). Furthermore, PFK-1 is regulated by phosphorylation (Cai *et al.*, 1997;Kemp *et al.*, 1981). PFK-1 phosphorylation modulates affinity for allosteric regulators as well as translocation between cytosol and cytoskeleton (Cai *et al.*, 1997;Kitajima *et al.*, 1983;Luther & Lee, 1986). Although PFK-1 has been thoroughly studied, its complex regulation is far from understood.

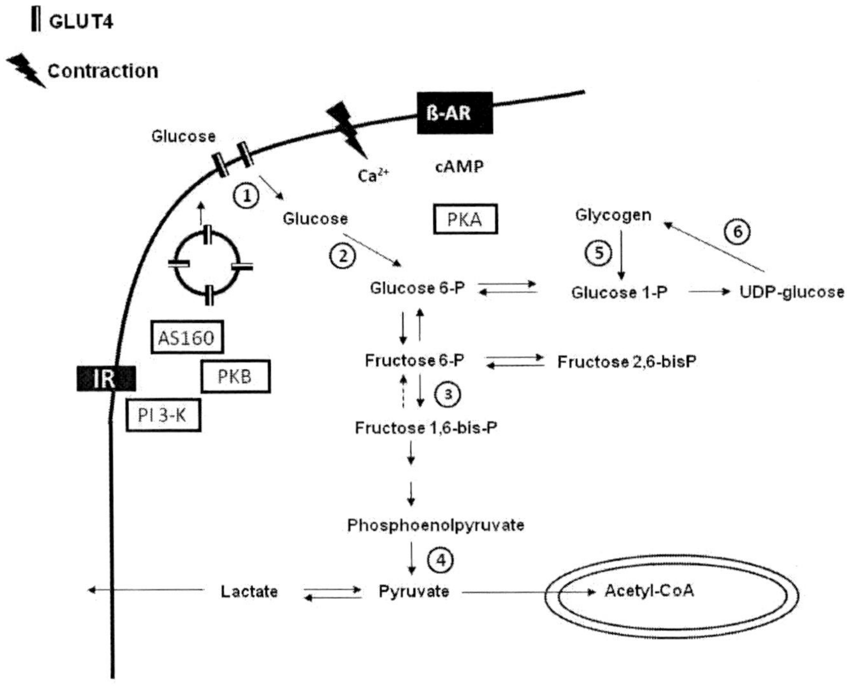

Figure 1. Schematic overview of regulation of glycolysis in skeletal muscles. The steps regulating glycolytic flux under different stimuli are marked as circled numbers. The regulating steps are: 1. Glucose transport is regulated by translocation of GLUT4 (Insulin and contraction stimulate translocation); 2) glucose phosphorylation by hexokinase is inhibited by glucose 6-phosphate (adrenaline-mediated increase in glucose 6-phosphate inhibits glucose phosphorylation); 3) PFK-1 activity is the key regulator of glycolytic flux (See text for further information); 4) pyruvate kinase activity seems not to be a regulating step in skeletal muscles; 5) Glycogen phosphorylase supplies glycolysis with glucose 1-phosphate (Adrenaline and contraction increases glycogenolysis); 6) Glycogen synthase activity determines glucose handling in skeletal muscles and therefore indirectly regulate glycolytic flux (Insulin activates glycolysis and channels glucose into glycogen synthesis). Abbreviations: β-AR: β-adrenergic receptor, AS160: Akt substrate of 160 kDa; GLUT4: glucose transporter 4, IR: insulin receptor, PI 3-K: phosphatidylinositol 3-kinase, PKA: protein kinase A (cAMP-dependent kinase), PKB: protein kinase B (Akt).

In 1980 van Schaftingen, Hue and Hers reported the existence of fructose 2,6-bisphosphate (van Schaftingen *et al.*, 1980). It became soon clear that fructose 2,6-bisphosphate is the key activator of phosphofructokinase-1 and regulator of glycolysis (Hers & Hue, 1983). Fructose 2,6-bisphosphate is produced from fructose 6-phosphate by 6-phosphofructo-2-kinase/fructose-2,6-bisphosphatase (PFK-2). PFK-2 is a bifunctional enzyme and also converts fructose 2,6-phosphate back to fructose 6-phosphate (Rider *et al.*, 2004). Three genes code PFK-2 and some splice variants exists; the skeletal muscle isoform is a splice variant of the gene coding the liver isoform (see below). PFK-2 is an unusual enzyme because the protein has two catalytic sites catalysing the opposite reactions. However, the two sites with opposite activities in the same protein makes it possible to co-ordinately regulate both synthesis and degradation of fructose 2,6-bisphosphate by a single signal.

Pyruvate kinase (PK) catalyses the final step of glycolysis where phosphoenolpyruvate is converted to pyruvate with formation of ATP. Pyruvate kinase has two genes (PKL and PKM) both of which give rise to two isoforms by alternative splicing. The pyruvate kinase reaction is irreversible, and other metabolic reactions are required for gluconeogenesis to proceed. For gluconeogenesis, pyruvate is transported into the mitochondria and transformed to oxaloacetate; oxaloacetate returns to the cytosol where PEPCK transforms it to phosphoenolpyruvate. Transformation of two phosphoenolpyruvate molecules to glucose 6-phosphate occurs through the reverse reaction of glycolysis but fructose 1,6 bisphosphatase bypasses the irreversible PFK-1 reaction. For efficient gluconeogenesis to occur inhibition of pyruvate kinase activity is necessary. The liver isoform of pyruvate kinase is phosphorylated by PKA (inhibiting activity) during glucagon-stimulated gluconeogenesis.

Glucose is transported into all cell type by transport proteins. Transport of glucose into cells is mediated by 13 different glucose transporters with tissue specific expression and different kinetic properties (Scheepers *et al.*, 2004). In some cell types (e.g. liver and pancreatic β-cell), glucose transport is only determined by concentration of extracellular glucose. Other cell types (like skeletal muscle) regulates glucose uptake by translocation of glucose transporters (GLUT4) to the cell membrane where they transport glucose into the cell. Glucose uptake in muscles is regulated by stimuli that translocate GLUT4 to the membrane and the concentration of extracellular glucose concentration is of minor importance.

Glucose is phosphorylated to glucose 6-phosphate by hexokinase. Hexokinase exists in four different isoforms with different tissue expression and kinetic properties (Wilson, 2003). Hexokinases are named hexokinase I to IV; hexokinase IV is better known as glucokinase. Hexokinase I-III is inhibited by glucose 6-phosphate. Glucokinase is not inhibited by glucose 6-phosphate, which allows high glucose phosphorylation for glycogen synthesis or glycolysis (Agius, 2008). Glucokinase is mainly expressed in liver and pancreatic β-cells. Skeletal muscles express mainly hexokinase II (Wilson, 2003).

Glycogen content varies great between cell types with the highest concentration in the liver; skeletal muscles keep second place. Glycogen is broken down by glycogen phosphorylase. Glycogen phosphorylase exists in three isoforms; muscle, liver and brain isoforms. Glycogen phosphorylase is activated by phosphorylation of a single amino acid (Ser^{14}) by phosphorylase kinase and allosteric activated by AMP. Phosphorylase becomes activated by PKA-mediated phosphorylation of phosphorylase kinase. PKA is activated by glucagon in the liver and adrenaline in muscles. Exercise activates glycogen phosphorylase via Ca^{2+}-mediated activation of phosphorylase kinase.

The majority of glucose is metabolised via glycolysis, but small amounts enter the hexosamine biosynthetic pathway and the pentose shunt. Increased flux via the hexosamine pathway has been associated with insulin resistance (Buse, 2006). Interestingly, the hexosamine pathway produces UDP-glucosamine which is the substrate for reversal O-linked glycosylation of proteins (Copeland *et al.*, 2008). Proteins are O-linked glycosylated at serines and threonines, and the serines/threonines that become glycosylated are often the same as become phosphorylated but the effect is different (Copeland *et al.*, 2008).

Metabolism of glucose via the pentose shunt participates in lipid synthesis by supply of NADPH. However, the pentose pathway intermediate xylulose 5-phosphate is also an interesting molecule because it can activate PP2A and regulate gene expression via ChREBP (Ilzuka & Horikawa, 2008). Yet, ChREBP-mediated gene regulation has not been described in skeletal muscles. Regulation of glycolysis will channel glucose into the other metabolic pathways and has, therefore, other functions than provideing energy.

Glycolysis in Skeletal Muscles

The biochemical reactions for glycolysis are of course the same in skeletal muscles as for all other tissues. However, the enzymes that catalyse the rate-limiting reactions have different isoforms with tissue specific expression. Below, the isoform specific expression and regulation of the enzymes mediating rate-limiting steps in skeletal muscles will be described.

The main glucose transporter expressed in skeletal muscles is GLUT4 (the insulin-regulated glucose transporter) which is regulated by translocation (James *et al.*, 1988). GLUT4 is stored in vesicles inside the cells and is translocated to the membrane upon stimulation by insulin or exercise (Etgen *et al.*, 1996). Rate of glucose transport seems to be regulated by the number of GLUT4 in the membrane (Etgen *et al.*, 1996). The two strongest stimulators of glucose uptake in skeletal muscles are insulin and exercise (Aslesen & Jensen, 1998;Etgen *et al.*, 1996;Jensen *et al.*, 1997). GLUT4 is also expressed in heart and adipocytes; two other insulin sensitive tissues.

Hexokinase exists in four different isoforms and skeletal muscles express mainly hexokinase II (Wilson, 2003). Hexokinase II activity is strongly inhibited by glucose 6-phosphate (Wilson, 2003). Hexokinase is normally not regarded as a rate-limiting step, but mice overexpressing hexokinase II do show higher insulin-stimulated glucose uptake (Chang *et al.*, 1996). Overexpression of hexokinase II in mice skeletal muscles have also been reported to increase performance during endurance exercise suggesting glucose phosphorylation as a rate-limiting step (Fueger *et al.*, 2005). During adrenaline stimulation, hexokinase become the rate-limiting step in skeletal muscles because hexokinase becomes inhibited by elevated concentration of glucose 6-phosphate (Aslesen & Jensen, 1998).

PFK-1 is considered as the enzyme that regulates glycolysis in skeletal muscles where glycolytic flux can increases more than 100-fold during extensive exercise (Connett & Sahlin, 1996). PFK-1 is highly regulated by numerous allosteric activators and PFK-1 has several binding sites for allosteric regulators (Cai *et al.*, 1997;Kemp & Foe, 1983). Concentration of metabolites like ADP, AMP, fructose 2,6-bisphosphate and P_i vary largely in skeletal muscles during e.g. exercise or hormonal stimulation. Skeletal muscle PFK-1 is also phosphorylated by PKA (Kemp *et al.*, 1981) but the physiological role of PFK-1 phosphorylation is not well documented (Kemp & Foe, 1983). However, PKA-mediated phosphorylation of muscle PFK-1 makes the enzyme more sensitive for inhibition by ATP and citrate (Foe & Kemp, 1982;Kitajima *et al.*, 1983). PFK-1 also translocates between cytosol and structural proteins (actin) and it has been reported that muscle contraction increases PFK-1 binding to myofibrillar proteins which prevents ATP-mediated inhibition whereas activation by both glucose 1,6-bisphosphate and fructose 2,6-bisphosphate was maintained (Andres *et al.*, 1996). This could be an important mechanism to maintain high glycolytic flux during exercise and still maintain high concentration of ATP for contractile activity. However, translocation of PFK-1, and other glycolytic enzymes to the contractile compartment, may also serve as a mechanism to produce ATP close to its utilization. Recently, it has also been reported that lactate reduces PFK-1 activity by favouring dissociation of the tetrameric complex (Costa Leite *et al.*, 2007). Interestingly, PKA as well as actin prevents lactate-mediated inhibition of PFK-1 activity (Costa Leite *et al.*, 2007) and the physiological role becomes therefore less clear. Muscles are, however, the specialised tissue for movement and intracellular regulation of PFK-1 by energy requirement and allosteric regulators seems logical.

As discussed above, fructose 2,6 bisphosphate is an important regulator of PFK-1 activity. The isoform of PFK-2 expressed in skeletal muscle is a splice variant of the liver isoform where the first 32 amino-acids are replaced by an unrelated nonapeptide (Crepin *et al.*, 1992). The skeletal isoform of PFK-2 lacks the PKA phosphorylation site and regulation of skeletal muscles PFK-2 activity has not been described. In fact, regulation of flux through PFK-2 may only be performed by substrate availability in skeletal muscles.

Pyruvate kinase has two genes (PKL and PKM) each giving rise to two splice variants. Skeletal muscles express the M_1-pyruvate kinase (Yamada & Noguchi, 1999a). The isoform of pyruvate kinase expressed is skeletal muscle is, in contrasts to the other pyruvate kinase isoforms, not regulated allosterically (Ikeda *et al.*, 2000). M_1-pyruvate kinase is not phosphorylated by PKA (Yamada & Noguchi, 1999b). So it is uncertain whether the M_1-pyruvate kinase in skeletal muscles is regulated at all. Inhibition of pyruvate kinase favours gluconeogenesis and liver pyruvate kinase is phosphorylated and inhibited by PKA during glucagon stimulation. In skeletal muscles, gluconeogenesis is not of major importance in skeletal muscles but glycogen can be synthesised from lactate.

Glycogen phosphorylase has been thoroughly studied in skeletal muscles, and activity is regulated by phosphorylation and allosterically activated by AMP. Other allosteric regulators exist and IMP increases phosphorylase activity whereas glucose 6-phosphate and low concentration of P_i decreases glycogen phosphorylase activity (Connett & Sahlin, 1996;Newsholme & Leech, 1983). Adrenaline increases glycogen phosphorylase activity by phosphorylation of Ser^{14} (See section adrenaline). Exercise also increases glycogen phosphorylase activity via Ca^{2+}-mediated activation of phosphorylase kinase.

Importantly, glycolysis requires NAD^+ in the reaction where glyceralde 3-phosphate is converted to 1,3-bisphosphoglycerate. The produced NADH can be oxidised in the mitochondria and regenerate NAD^+. However, skeletal muscles have a much larger capacity for glycolytic flux than for oxidative phosphorylation. During short-term high intensity exercise, NAD^+ can be regenerated by transformation of pyruvate to lactate. Lactate dehydrogenase catalyses this reaction, which couples conversion of pyruvate to $NADH/NAD^+$ transformation. The regenerated NAD^+ allows glycolysis to continue; however, the accumulation of lactate will rapidly cause acidosis. Of note, different skeletal muscle fibre type exists. These fibre types are classified as slow-twitch oxidative (type I), fast-twitch oxidative (type IIA) and fast-twitch glycolytic fibres (Pette & Staron, 1990). Glycolytic capacity varies greatly between type I and type II fibres, but all fibre types expresses the same isoforms of the rate-limiting glycolytic enzymes, and fibre type differences will not be discussed.

Lactate is transported out of the muscle cells by lactate transportes (MCT1 and MCT4) and becomes available to other cells where lactate can be oxidized or transformed to glucose via gluconeogenesis (Gladden, 2004). Glycolysis in skeletal muscles is therefore an integrated part of regulation of whole body carbohydrate metabolism, and lactate is an important intermediate for energy transfer between tissues.

Insulin-Mediated Activation of Glycolysis

Insulin has profound effects on glucose metabolism in skeletal muscles and 70-90 % of insulin-mediated glucose disposal occurs in muscles (Shulman *et al.*, 1990). The major part of the glucose taken up is incorporated into glycogen (Shulman *et al.*, 1990). Glycogen synthase, which incorporates the glucose moieties into the glycogen particle, represents a rate-limiting step for glycogen synthesis and is activated by insulin (Cohen, 1993;Jensen *et al.*, 2006). However, part of the glucose taken up is metabolised directly via glycolysis and insulin increases glycolytic flux (Dimitriadis *et al.*, 1992;Jensen *et al.*, 2006).

Insulin stimulates glucose uptake via translocation of GLUT4 from intracellular vesicles to the cell membrane (James *et al.*, 1988). The signalling pathway has to some degree been resolved. The insulin receptor is a tyrosine kinase, which phosphorylates the receptor and Insulin Receptor Substrate-1 (IRS-1). IRS-1 binds and activates PI 3-kinase and the lipid kinase increases PIP_3 in the membrane which activates PKB (Shepherd, 2005). Activated PKB phosphorylates AS160 for translocation of GLUT4 (Ramm *et al.*, 2006). The signalling molecules downstream of AS160 remains to be determined but insulin stimulates translocation of GLUT4 to the cell membrane and transport of glucose into muscle cells.

Glucose phosphorylation (by hexokinase II) may be a rate-limiting step for insulin-mediated glucose utilization (Chang *et al.*, 1996), but it has been difficult to show any significant accumulation of free glucose unless adrenaline is present (Aslesen & Jensen, 1998). However, glucose phosphorylation becomes the rate-limiting step when insulin and adrenaline are present simultaneously. Adrenaline does not reduce insulin-stimulated glucose transport (GLUT4 translocation) but increases glucose 6-phosphate concentration and blocks glucose phosphorylation (Aslesen & Jensen, 1998).

Glycogen synthase is regulated by phosphorylation and allosterically activated by glucose 6-phosphate (Cohen, 1993). GSK-3 phosphorylates glycogen synthase at several sites which reduces activity. Insulin activates glycogen synthase via PI 3-kinase, PKB and GSK-3; PKB phosphorylates and inactivates GSK-3 which decreases glycogen synthase phosphorylation and increases activity (McManus *et al.*, 2005). Intracellular content of glucose 6-phosphate increases in skeletal muscles during insulin stimulation which contributes to stimulation of glycogen synthesis (Bouskila *et al.*, 2008; Dimitriadis *et al.*, 1992; Jensen *et al.*, 2006;Shulman *et al.*, 1990). During insulin stimulation glucose can either be channelled to glycogen synthesis or to glycolysis. The glycogen content influences glucose handling in skeletal muscles during insulin stimulation (Jensen *et al.*, 2006). In muscles with normal glycogen content, we find that about 90 % of the glucose is incorporated into glycogen and on 5-10 % passes through glycolysis (Jensen *et al.*, 2006). Although rate of glucose uptake is much higher in muscles with low glycogen content high glycogen synthase activation channels more than 95 % of the glucose into glycogen and glycolytic flux is not increased. In muscles with high glycogen content, glycogen synthase activity is low and glycolysis higher during insulin stimulation than in muscles with normal glycogen content (Jensen *et al.*, 2006). Glycogen synthase activity, therefore, indirectly regulate glycolytic flux in skeletal muscles.

Glucose 6-phosphate is converted to fructose 6-phosphate by phosphoglucose isomerase in an equilibrium reaction. High concentration of glucose 6-phosphate will therefore increase fructose 6-phosphate concentration and substrate availability for PFK-1. Furthermore, insulin increases content of fructose 2,6-bisphosphate in skeletal muscles (Dimitriadis *et al.*, 1992) which will activate PFK-1. Insulin has not been reported to activate PFK-2 by phosphorylation in skeletal muscles. However, it has been suggested that insulin translocates PFK-1 to the myofibrillar fraction and increase activity (Silva *et al.*, 2004), but the experiments were performed in muscle homogenates and physiological activation of PFK-1 by insulin has not been reported.

As mentioned above, glycolytic flux was increased in muscles with high glycogen content. In muscles high glycogen insulin also increases glucose 6-phosphate to higher levels and glycolysis is higher than in muscles with normal or low glycogen content. Increased substrate availability may be sufficient to increase glycolysis during insulin stimulation (Jensen *et al.*, 2006). Although insulin stimulation increases glycolysis and lactate release from skeletal (Dimitriadis *et al.*, 1992;Jensen *et al.*, 2006;Qvisth *et al.*, 2007), we have not been able to find significant elevation in lactate concentration in skeletal muscles during insulin stimulation (Aslesen & Jensen, 1998), which suggest that lactate is rapidly transported out of muscle cells.

Insulin seems to activate glycolysis in skeletal muscles solely by increasing substrate supply. However, muscle glycogen content and glycogen synthase activity determine the fraction that is channeled into glycolysis.

Adrenaline-Mediated Activation of Glycolysis

Adrenaline is a strong activator of glycogen phosphorylase and stimulates glycogen breakdown (Jensen & Dahl, 1995). Adrenaline activates glycogen phosphorylase via β-adrenergic receptors, cAMP and activation of PKA. Activated PKA phosphorylates phosphorylase kinase, which phosphorylates glycogen phosphorylase Ser^{14} and activates the enzyme. During adrenaline stimulation, glycogen is broken down, glycolysis increased and lactate accumulates in muscles (Aslesen & Jensen, 1998;Challiss *et al.*, 1986;Jensen & Dahl, 1995).

Adrenaline stimulates glycolysis in skeletal muscles (Challiss *et al.*, 1986). However, the adrenaline-mediated glycogen breakdown increases glucose 6-phosphate to high levels in skeletal muscles which blocks hexokinase activity and therefore glucose phosphorylation (Aslesen & Jensen, 1998;Wilson, 2003). Indeed, adrenaline only increases glycolysis from glycogen whereas lactate formation from external glucose is unchanged (Challiss *et al.*, 1986). This is in contrast to insulin stimulation where glycogen accumulates and glycolysis is fed by external glucose.

Glycogen phosphorylase is the enzyme that regulates glycogenolysis, and this step is important for adrenaline-mediated increase in glycolysis. However, although adrenaline is a strong activator of glycogen phosphorylase and glycogen breakdown, glycolytic flux is limited during adrenaline stimulation compared to muscle contraction (Aslesen & Jensen, 1998;Jensen & Dahl, 1995). Furthermore, adrenaline increases concentration of glucose 6-phosphate (Aslesen & Jensen, 1998) and fructose 2,6-bisphosphate (Jones *et al.*, 1994), which should stimulate glycolysis. However, PFK-1 is regulated by numerous of allosteric regulators and glycolysis cannot run independent of energy utilisation. Adrenaline increases metabolic rate in skeletal muscles by about 50 % (Simonsen *et al.*, 1999).

Adrenaline increases PFK-1 phosphorylation which has been reported to increase ATP-mediated inhibition (Foe & Kemp, 1982). Therefore, increased PFK-1 phosphorylation (and affinity for ATP-mediated inhibition) may counteract the expected activation of glycolytic flux from the increase concentration of substrate and the allosteric activator fructose 2,6-bisphosphate. Adrenaline stimulation also increases the intracellular concentration of lactate and lactate has been reported to decrease PFK-1 activity (Costa Leite *et al.*, 2007). However, PKA-mediated phosphorylation prevented lactate-mediated inhibition of PFK-1. The mechanisms controlling glycolytic flux in skeletal muscles during adrenaline stimulation are complex and poorly understood.

Skeletal muscles are not able to release glucose into the blood stream as the liver does. The reason is that skeletal muscles lack glucose 6-phosphatase activity. However, skeletal muscles can release lactate which can be converted to glucose via gluconeogenesis in the liver or renals (Meyer *et al.*, 2004). The lactate released can also serves as a substrate for the heart. Skeletal muscles contain the major part ($\approx$ 80%) of the body's carbohydrates and breakdown of glycogen to lactate makes the carbohydrate source in skeletal muscles available for other tissues. In fact, lactate flux is substantial and provides an important interchange of carbohydrate source between tissues.

Adrenaline seems to activate glycolysis in skeletal muscles mainly by activation of glycogen phosphorylase and increasing substrate supply. However, PKA-mediated phosphorylation and allosteric regulators control (limits) glycolytic flux. Adrenaline-stimulated glycolysis in skeletal muscles makes muscle glycogen available for other tissues (in the form of lactate) and muscle glycolysis is important for whole body metabolic regulation.

Contraction-Mediated Activation of Glycolysis

Skeletal muscle cells are specialised for movement and have a high capacity for converting chemical energy into mechanical work. In skeletal muscles energy production has priority, although fatigue mechanisms protect against irreversible damage. During muscle contraction, glycolysis increases dramatically and short-sprint exercise can increase by 100-fold (Connett & Sahlin, 1996). The intensity of exercise determines glycolytic rate and skeletal muscles seem perfectly to adjust glycolytic flux to demands during exercise (Connett & Sahlin, 1996).

Energy status of muscle cells is a major determinant of glycolytic flux and complex allosteric regulation prevents significant decline in ATP. Glycolytic energy production is particularly important during high intensity exercise (above 100 % of maximal oxygen consumption) for relatively short periods of time (less than 10 min). During aerobic conditions, the mitochondria will oxidize the NADH produced in glycolysis and regenerate NAD^+. During high intensity exercise NAD^+ can be regenerated by transformation of pyruvate to lactate, as lactate dehydrogenase will oxidize NADH to NAD^+ when pyruvate is transformed to lactate. However, abundant production of lactate will decrease pH and cause fatigue.

Glycogen is the important carbohydrate during exercise. Exercise increases glycogen phosphorylase activity and stimulates glycogen breakdown (Aslesen & Jensen, 1998;Hespel & Richter, 1992). Exercise rapidly increases glycogen phosphorylase in the a-form (phosphorylated), but percent phosphorylase a return to basal level in less than 10 min despite that glycogenolysis remains high. After the first minutes of exercise glycogenolysis is activated allosterically. Exercise, therefore, activates phosphorylase by phosphorylation as well as allosteric mechanisms.

Exercise increases content of glucose 6-phosphate. Exercise also increases fructose 2,6-bisphosphate content in skeletal muscles which will stimulate glycolysis (Winder & Duan, 1992). Chronic stimulation of muscles increases fructose-2,6-bisphosphate in muscles which increases glycolysis (Cadefau *et al.*, 1999) and PFK-1 is activated allosterically during exercise. Furthermore, exercise has been reported to translocate PFK-1 to myofibrillar proteins (Andres *et al.*, 1996). PFK-1 association with myofibrils prevents ATP-mediated inhibition while glucose 1,6-bisphosphate and fructose 2,6-bisphosphate still activates PFK-1 (Andres *et al.*, 1996). Contraction has also been reported to increase PFK-1 phosphorylation, but a physiological role of this phosphorylation is uncertain.

Exercise stimulates glucose uptake and metabolism of glucose. Glycogen content influences contraction-stimulated glucose uptake and contraction-stimulated glucose uptake is higher in muscles with low glycogen (Derave *et al.*, 1999) and (Lai and Jensen, unpublished). During exercise, plasma adrenaline increases (Kjær *et al.*, 1987) and adrenaline increases glucose 6-phosphate concentration in contracting muscles to levels expected to inhibits hexokinase activity (Aslesen & Jensen, 1998). Interestingly, glucose phosphorylation occurs in contracting muscles even with high concentrations of glucose 6-phosphate (Aslesen & Jensen, 1998). The mechanism has not been ruled out, but hexokinase II binds to the mitochondria which changes regulation (Wilson, 2003). Importantly, the mechanisms ensure that exercising muscles can use blood glucose whereas adrenaline completely blocks glucose utilization in non-active muscles. This channels glucose to active muscles and may be an important survival mechanism in the "fight or flight" reaction.

Contraction activates glycolysis in skeletal muscles by increasing substrate supply and modulating PFK-1 allosteric regulation. During contraction, PFK-1 translocates to myofibrils which ensure high enzyme activity concomitantly with high ATP content for contraction. Furthermore, allosteric activators increase glycolytic flux. The fact that muscles are the specialised tissue for movement with rapid changes in energy requirements supports an intracellular allosteric regulation by energy status.

Regulation of Glycolysis by Fat (Randle Cycle)

Carbohydrates are normally metabolised together with lipids and the ratio is regulated by many factors including diet, exercise and training status. The glucose-fatty acid cycle indicates that increased availability of FFA decreases glucose utilisation. The glucose-fatty acid cycle states that high amount of FFA available for β-oxidation will increase concentration of acyl-CoA and citrate. Acyl-CoA will inhibit PDH and therefore pyruvate oxidation while citrate will inhibit PFK-1. Inhibition of PFK-1 will favour accumulation of glucose 6-phodphate and high concentration of glucose 6-phosphate will inhibit hexokinase and therefore glucose uptake (Newsholme & Leech, 1983).

Importantly, Randle and colleagues did the majority of experiments on the heart to describe the glucose-fatty acid cycle and glycolysis is differently regulated heart and muscle (Depre *et al.*, 1998). It has also been difficult to show the glucose-fatty acid cycle in skeletal muscles, although carbohydrate and lipid metabolism also have interaction in skeletal muscles (Spriet & Watt, 2003).

Ingestion of carbohydrates increases carbohydrate oxidation and decreases fat oxidation rather the other way around (Flatt, 1995;Woerle *et al.*, 2003). The human body has limited capacity to store carbohydrate and excess carbohydrates have to be utilised or converted to lipid when glycogen stores are filled. Mostly, at least 50 % of the energy intake is carbohydrate and this shows that the body has a large capacity to remove glucose from the blood. Lipid synthesis from glucose requires that glucose pass through glycolysis. It is likely that skeletal muscles will glycolyse a large part of the glucose glycolysis and release lactate for lipid synthesis in liver. High FFA does not seem to prevent glycolysis in skeletal muscles

Glycolysis and Diabetes

Obesity and type 2 can be viewed as an "energy over-supply syndrome" where glucose and lipid are in excess. Glucose oxidation is normally higher in type 2 diabetics than controls in basal conditions (Højlund *et al.*, 2008). However, insulin increases glucose oxidation in normal subjects whereas insulin is unable to increase glucose oxidation in type 2 diabetics (Højlund *et al.*, 2006). Insulin's inability to increase glucose oxidation is called metabolic inflexibility, and skeletal muscles contribute to the increased glucose oxidation. For glucose oxidation, glucose has to be transported into cells, pass through glycolysis and enter mitochondria for oxidation. Therefore, the defect in diabetic muscles can be reduced insulin-stimulated glucose uptake, glycolytic capacity or reduced oxidative capacity.

It is well-documented that insulin-stimulated glucose uptake is reduced in skeletal muscles of type 2 diabetics (DeFronzo *et al.*, 1985). However, insulin-stimulated glycogen synthase activation is also decreased (Højlund *et al.*, 2006). As discussed previously, reduced glycogen synthase activation in muscles with high glycogen content channels more glucose into glycolysis. Whether insulin-stimulated glycolysis is higher in muscle of type 2 diabetics is uncertain. Several studies have reported higher lactate concentrations in type 2 diabetics (Reaven *et al.*, 1988) and skeletal muscles normally contribute with the majority of lactate (Consoli *et al.*, 1990). However, skeletal lactate release seems to be impaired in muscles of obese insulin resistant humans (Qvisth *et al.*, 2006).

Glycolytic capacity seems not to be reduced in type 2 diabetes, and higher LDH activity has been reported in human muscles from type 2 diabetics (Oberbach *et al.*, 2006). Type 2 diabetics do not have reduced PFK-1 activity (Simoneau & Kelley, 1997). Instead, the ratio between PFK-1 and citrate synthase activity is inversely correlated with insulin-stimulated glucose disposal (Simoneau & Kelley, 1997). Therefore, insulin resistance seem to be related to mitochondrial dysfunction rather than reduced glycolysis (Højlund *et al.*, 2008;Simoneau & Kelley, 1997). However, PFK-1 activity has been reported to be reduced in obese diabetic db/db mice (Bazaes *et al.*, 1982). Indeed glycolytic capacity is much higher than oxidative capacity in skeletal muscles, but the role of glycolysis in skeletal muscles for development of type 2 diabetes deserves more focus in future research.

Muscle atrophy is associated with metabolic disturbances. Furthermore, diabetes is associated with muscle atrophy. Muscle wasting is in part regulated by the ubiquitin ligase named muscle-specific RING-finger protein 1 (MuRF-1). However, muscle specific overexpression of MuRF-1 did not reduce body weight but caused changes in whole body carbohydrate metabolism including reduced liver glycogen content and hyperinsulinemia (Hirner *et al.*, 2008). Interestingly, MuRF-1 interact with multiple enzymes involved in glucose metabolism including pyruvate kinase and PDH (Hirner *et al.*, 2008). Overall these data show that changes in skeletal muscle metabolism causes whole body changes in metabolic regulation.

Unfortunately, genetic approaches have not been used to study the role of PFK-2 for metabolic regulation during insulin stimulation in skeletal muscles. Although the heart expresses another isoform of PFK-2, which is regulated by phosphorylation, expression of a kinase deficient PFK-2 reduced glycolysis in the heart (Donthi *et al.*, 2004). In the heart, glucose uptake is also regulated via translocation of GLUT4. Interestingly, heart cell expressing kinase dead PFK-2 had increased glycogen content and was insulin resistant (Donthi *et al.*, 2004).

Increasing liver glycolysis by adenovirus-mediated overexpression of glucokinase or PFK-2 increased glycolysis and decreased blood glucose and insulin (Wu *et al.*, 2005). Furthermore, hepatic glucose output decreased and insulin-mediated glucose disposal increased. Interestingly, skeletal muscle capacity for palmitate oxidation increased, most likely because expression of ACC1 and FAS decreased (Wu *et al.*, 2005). These data suggest that decreasing availability of blood glucose for muscle glycolysis is beneficial for skeletal β-oxidative capacity; endurance training has the same effect. We suggest that excess glycolysis in skeletal muscles cause accumulations of metabolites which deteriorates insulin sensitivity in skeletal muscles. Whether this is increased flux via hexosamine pathway, accumulation of xylose 5-phosphate, citrate (for lipid synthesis) or other metabolites remains to be determined.

Muscles from diabetics have reduced oxidative capacity (Mogensen *et al.*, 2007;Petersen *et al.*, 2003). Indeed low aerobic fitness and oxidative capacity is a hallmark of type 2 diabetics. Endurance training increases insulin sensitivity and increased mitochondrial function seems part of the beneficial effect of training. The other beneficial effect of endurance training is that muscle glycogen is used. Muscles with low glycogen efficiently stores glucose as glycogen (Jensen *et al.*, 2006) which improve whole body glucose regulation. Reduced skeletal muscle glycolytic capacity does not limit glucose metabolism.

Human do have high capacity to convert glucose into lipid when the glycogen stores are filled and a high caloric carbohydrate diet is consumed (Acheson *et al.*, 1988). It is commonly believed that the majority of lipid synthesis from glucose occurs in liver cells. However, skeletal muscle can synthesise lipid from glucose (Aas *et al.*, 2004). We were the first to describe that incubation human primary muscle cell with high glucose resulted in lipid accumulation rather than increasing glycogen content (Aas *et al.*, 2004) and insulin resistant skeletal muscles are characterised with high content of triacylglycerol (Franch *et al.*, 2002). However, insulin resistant muscles seem to maintain its glycolytic capacity.

In type 2 diabetes, glycogen cycling is increased (Woerle *et al.*, 2006). Glycogen cycling normally requires that glucose is metabolised via both glycolysis and gluconeogenesis. Gluconeogenesis occurs mainly in liver and renals, whereas a major part of glycolysis occurs in skeletal muscles. I hypothesise that high glycogen in muscles prevents the normal storage of glucose as muscle glycogen content and increases glycolysis and lactate release. Furthermore, when the glycogen stores are filled carbohydrates will continue to be cycled in the Cori cycle until the energy is used or converted to lipid.

Conclusions

Skeletal muscles store the majority of the body's carbohydrates and play an important role in whole body glucose metabolism. When glucose is inside muscle cell it must be metabolised via glycolysis and completely oxidised or released as lactate. Insulin increases skeletal muscle glucose uptake and glycolysis, but insulin also activates glycogen synthase and channels the majority of glucose into glycogen. However, when glycogen content is high, insulin-stimulated glycolysis will increase and some of the metabolic intermediates may over time cause insulin resistance. Adrenaline stimulates glycogen breakdown and increases glycolytic flux. However, adrenaline also increases glucose 6-phosphate concentration to high levels which blocks hexokinase activity and prevents glucose phosphorylation and therefore prevent that blood glucose enters glycolysis. Exercise is the strongest activator of glycolysis in skeletal muscles; the substrate is mainly glycogen but exercise also stimulates glucose uptake. Interestingly, adrenaline does not prevent that skeletal muscles use blood glucose during exercise since glucose 6-phosphat is a poor inhibitor of hexokinase activity in contracting muscles. Skeletal muscles play a key role in regulation of whole body carbohydrate metabolism and skeletal muscle glycolysis is an important pathway for metabolic regulation. Increased glycolytic flux through non-contracting skeletal muscle may cause insulin resistance.

References

Aas V, Kase ET, Solberg R, Jensen J, & Rustan AC (2004). Chronic hyperglycaemia promotes lipogenesis and triacylglycerol accumulation in human skeletal muscle cells. *Diabetologia.* 47, 1452-1461.

Acheson KJ, Schutz Y, Bessard T, Anantharaman K, Flatt JP, & Jequier E (1988). Glycogen storage capacity and de novo lipogenesis during massive carbohydrate overfeeding in man. *Am. J. Clin. Nutr.* 48, 240-247.

Agius L (2008). Glucokinase and molecular aspects of liver glycogen metabolism. *Biochem. J.* 414, 1-18.

Andres V, Carreras J, & Cusso R (1996). Myofibril-bound muscle phosphofructokinase is less sensitive to inhibition by ATP than the free enzyme, but retains its sensitivity to stimulation by bisphosphorylated hexoses. *Int. J. Biochem. Cell Biol.* 28, 1179-1184.

Aslesen R & Jensen J (1998). Effects of epinephrine on glucose metabolism in contracting rat skeletal muscle. *Am. J. Physiol.* 275, E448-E456.

Bazaes SE, Foe LG, & Kemp RG (1982). Phosphate content of muscle phosphofructokinase in the genetically diabetic mouse (C57BL/KsJ). *Arch. Biochem. Biophys.* 218, 483-487.

Bouskila M, Hirshman MF, Jensen J, Goodyear LJ, & Sakamoto K (2008). Insulin promotes glycogen synthesis in the absence of GSK3 phosphorylation in skeletal muscle. *Am. J. Physiol. Endocrinol. Metab.* 294, E28-E35.

Buse MG (2006). Hexosamines, insulin resistance, and the complications of diabetes: current status. *Am. J. Physiol. Endocrinol. Metab.* 290, E1-E8.

Cadefau JA, Parra J, Tauler A, & Cusso R (1999). Contractile activity modifies Fru-2,6-P(2) metabolism in rabbit fast twitch skeletal muscle. *J. Biol. Chem.* 274, 31961-31966.

Cai GZ, Callaci TP, Luther MA, & Lee JC (1997). Regulation of rabbit muscle phosphofructokinase by phosphorylation. *Biophys. Chem.* 64, 199-209.

Challiss RAJ, Lozeman FJ, Leighton B, & Newsholme EA (1986). Effects of the b-adrenoceptor agonist isoprenaline on insulin-sensitivity in soleus muscle of the rat. *Biochem. J.* 233, 377-381.

Chang PY, Jensen J, Printz RL, Granner DK, Ivy JL, & Moller DE (1996). Overexpression of hexokinase II in transgenic mice - Evidence that increased phosphorylation augments muscle glucose uptake. *J. Biol. Chem.* 271, 14834-14839.

Cohen P. (1993). Dissection of the protein phosphorylation cascades involved in insulin and growth factor action. *Biochem. Soc. Trans.* 214, 555-567.

Connett RD & Sahlin K (1996). Control of glycolysis and glycogen metabolism. In *Handbook of Physiology. Integration of motor, circulatory, respiratory and metabolic control during exercise.*, eds. Rowell LB & Shepherd JT, pp. 870-911. American Physiological Society.

Consoli A, Nurjhan N, Reilly JJ, Jr., Bier DM, & Gerich JE (1990). Contribution of liver and skeletal muscle to alanine and lactate metabolism in humans. *Am. J. Physiol.* 259, E677-E684.

Copeland RJ, Bullen JW, & Hart GW (2008). Cross-talk between GlcNAcylation and phosphorylation: roles in insulin resistance and glucose toxicity. *Am. J. Physiol. Endocrinol. Metab.* 295, E17-E28.

Cori CF & Cori GT (1928). The mechanism of epinephrine action. I. The influence of epinephrine on the carbohydrate metabolism of fasting rats, with a note on new formation of carbohydrates. *J. Biol. Chem.* 79, 309-319.

Costa Leite T, Da Silva D., Guimaraes Coelho R., Zancan P, & Sola-Penna M (2007). Lactate favours the dissociation of skeletal muscle 6-phosphofructo-1-kinase tetramers down-regulating the enzyme and muscle glycolysis. *Biochem. J.* 408, 123-130.

Coyle EF (1995). Substrate utilization during exercise in active people. *Am. J. Clin. Nutr.* 61, 968S-979S.

Crepin KM, De CM, Vertommen D, Foret D, Michel A, Rider MH, Rousseau GG, & Hue L (1992). Molecular forms of 6-phosphofructo-2-kinase/fructose-2,6-bisphosphatase expressed in rat skeletal muscle. *J. Biol. Chem.* 267, 21698-21704.

DeFronzo RA, Gunnarsson R, Björkman O, Olsson M, & Wahren J (1985). Effect of insulin on peripheral and splanchnic glucose metabolism in noninsulin-dependent (typeII) diabetes mellitus. *J. Clin. Invest.* 76, 149-155.

Depre C, Rider MH, & Hue L (1998). Mechanisms of control of heart glycolysis. *Eur. J. Biochem.* 258, 277-290.

Derave W, Lund S, Holman GD, Wojtaszewski J, Pedersen O, & Richter EA (1999). Contraction-stimulated muscle glucose transport and GLUT-4 surface content are dependent on glycogen concentration. *Am. J. Physiol.* 277, E1103-E1110.

Dimitriadis G, Parrybillings M, Bevan S, Dunger D, Piva T, Krause U, Wegener G, & Newsholme EA (1992). Effects of Insulin-Like Growth Factor-I on the Rates of Glucose Transport and Utilization in Rat Skeletal Muscle Invitro. *Biochem. J.* 285, 269-274.

Donthi RV, Ye G, Wu C, McClain DA, Lange AJ, & Epstein PN (2004). Cardiac expression of kinase-deficient 6-phosphofructo-2-kinase/fructose-2,6-bisphosphatase inhibits glycolysis, promotes hypertrophy, impairs myocyte function, and reduces insulin sensitivity. *J. Biol. Chem.* 279, 48085-48090.

Etgen GJ, Wilson CM, Jensen J, Cushman SW, & Ivy JL (1996). Glucose transport and cell surface GLUT-4 protein in skeletal muscle of the obese Zucker rat. *Am. J. Physiol.* 271, E294-E301.

Flatt JP (1995). Use and storage of carbohydrate and fat. *Am. J. Clin. Nutr.* 61, S952-S959.

Foe LG & Kemp RG (1982). Properties of phospho and dephospho forms of muscle phosphofructokinase. *J. Biol. Chem.* 257, 6368-6372.

Franch J, Knudsen J, Ellis BA, Pedersen PK, Cooney GJ, & Jensen J (2002). Acyl-CoA binding protein expression is fibre type specific and elevated in muscles from obese insulin-resistant Zucker rat. *Diabetes.* 51, 449-454.

Fueger PT, Shearer J, Krueger TM, Posey KA, Bracy DP, Heikkinen S, Laakso M, Rottman JN, & Wasserman DH (2005). Hexokinase II protein content is a determinant of exercise endurance capacity in the mouse. *J. Physiol.* 566, 533-541.

Gladden LB (2004). Lactate metabolism: a new paradigm for the third millennium. *J. Physiol.* 558, 5-30.

Hers HG & Hue L (1983). Gluconeogenesis and related aspects of glycolysis. *Ann. Rev. Biochem.* 52, 617-653.

Hespel P & Richter EA (1992). Mechanism linking glycogen concentration and glycogenolytic rate in perfused contracting rat skeletal muscle. *Biochem. J.* 284, 777-780.

Hirner S, Krohne C, Schuster A, Hoffmann S, Witt S, Erber R, Sticht C, Gasch A, Labeit S, & Labeit D (2008). MuRF1-dependent regulation of systemic carbohydrate metabolism as revealed from transgenic mouse studies. *J. Mol. Biol.* 379, 666-677.

Højlund K, Frystyk J, Levin K, Flyvbjerg A, Wojtaszewski JF, & Beck-Nielsen H (2006). Reduced plasma adiponectin concentrations may contribute to impaired insulin activation of glycogen synthase in skeletal muscle of patients with type 2 diabetes. *Diabetologia.* 49, 1283-1291.

Højlund K, Mogensen M, Sahlin K, & Beck-Nielsen H (2008). Mitochondrial dysfunction in type 2 diabetes and obesity. *Endocrinol. Metab. Clin. North Am.* 37, 713-731.

Ikeda Y, Taniguchi N, & Noguchi T (2000). Dominant negative role of the glutamic acid residue conserved in the pyruvate kinase M(1) isozyme in the heterotropic allosteric effect involving fructose-1,6-bisphosphate. *J. Biol. Chem.* 275, 9150-9156.

Ilzuka K & Horikawa Y (2008). ChREBP: a glucose-activated transcription factor involved in the development of metabolic syndrome. *Endocr. J.* 55, 617-624.

James DE, Brown R, Navarro J, & Pilch PF (1988). Insulin-regulatable tissues express a unique insulin-sensitive glucose transport protein. *Nature.* 333, 183-185.

Jensen J, Aslesen R, Ivy JL, & Brørs O (1997). Role of glycogen concentration and epinephrine on glucose uptake in rat epitrochlearis muscle. *Am. J. Physiol.* 272, E649-E655.

Jensen J & Dahl HA (1995). Adrenaline stimulated glycogen breakdown in rat epitrochlearis muscles: Fibre type specificity and relation to phosphorylase transformation. *Biochem. Mol. Biol. Int.* 35, 145-154.

Jensen J, Jebens E, Brennesvik EO, Ruzzin J, Soos MA, Engebretsen EM, O'Rahilly S, & Whitehead JP (2006). Muscle glycogen inharmoniously regulates glycogen synthase activity, glucose uptake, and proximal insulin signaling. *Am. J. Physiol. Endocrinol. Metab.* 290, E154-E162.

Jones JP, Maclean PS, & Winder WW (1994). Correlation between fructose 2,6-bisphosphate and lactate production in skeletal muscle. *J. Appl. Physiol.* 76, 2169-2176.

Kemp RG & Foe LG (1983). Allosteric regulatory properties of muscle phosphofructokinase. *Mol. Cell Biochem.* 57, 147-154.

Kemp RG, Foe LG, Latshaw SP, Poorman RA, & Heinrikson RL (1981). Studies on the phosphorylation of muscle phosphofructokinase. *J. Biol. Chem.* 256, 7282-7286.

Kitajima S, Sakakibara R, & Uyeda K (1983). Significance of phosphorylation of phosphofructokinase. *J. Biol. Chem.* 258, 13292-13298.

Kjær M, Secher NH, & Galbo H (1987). Physical stress and catecholamine release. *Bailliere's Clin. Endocrin. Met.* 1, 279-298.

Luther MA & Lee JC (1986). The role of phosphorylation in the interaction of rabbit muscle phosphofructokinase with F-actin. *J. Biol. Chem.* 261, 1753-1759.

McManus EJ, Sakamoto K, Armit LJ, Ronaldson L, Shpiro N, Marquez R, & Alessi DR (2005). Role that phosphorylation of GSK3 plays in insulin and Wnt signalling defined by knockin analysis. *EMBO J.* 24, 1571-1583.

Meyer C, Woerle HJ, Dostou JM, Welle SL, & Gerich JE (2004). Abnormal renal, hepatic, and muscle glucose metabolism following glucose ingestion in type 2 diabetes. *Am. J. Physiol. Endocrinol. Metab.* 287, E1049-E1056.

Mogensen M, Sahlin K, Fernstrom M, Glintborg D, Vind BF, Beck-Nielsen H, & Højlund K (2007). Mitochondrial respiration is decreased in skeletal muscle of patients with type 2 diabetes. *Diabetes.* 56, 1592-1599.

Newsholme EA & Leech AR (1983). *Biochemistry for the medical sciences*, 1 ed., pp. 1-952. John Wiley & Sons, Chichester.

Oberbach A, Bossenz Y, Lehmann S, Niebauer J, Adams V, Paschke R, Schon MR, Bluher M, & Punkt K (2006). Altered fiber distribution and fiber-specific glycolytic and oxidative enzyme activity in skeletal muscle of patients with type 2 diabetes. *Diab. Care.* 29, 895-900.

Petersen KF, Befroy D, Dufour S, Dziura J, Ariyan C, Rothman DL, DiPietro L, Cline GW, & Shulman GI (2003). Mitochondrial dysfunction in the elderly: possible role in insulin resistance. *Science.* 300, 1140-1142.

Pette D & Staron RS (1990). Cellular and molecular diversities of mammalian skeletal muscle fibers. *Rev. Physiol. Biochem. Pharmacol.* 116, 1-76.

Qvisth V, Hagstrom-Toft E, Enoksson S, Moberg E, Arner P, & Bolinder J (2006). Human skeletal muscle lipolysis is more responsive to epinephrine than to norepinephrine stimulation in vivo. *J. Clin. Endocrinol. Metab.* 91, 665-670.

Qvisth V, Hagstrom-Toft E, Moberg E, Sjoberg S, & Bolinder J (2007). Lactate release from adipose tissue and skeletal muscle in vivo: defective insulin regulation in insulin-resistant obese women. *Am. J. Physiol. Endocrinol. Metab.* 292, E709-E714.

Ramm G, Larance M, Guilhaus M, & James DE (2006). A role for 14-3-3 in insulin-stimulated GLUT4 translocation through its interaction with the RabGAP AS160. *J. Biol. Chem.* 281, 29174-29180.

Reaven GM, Hollenbeck C, Jeng CY, Wu MS, & Chen YD (1988). Measurement of plasma glucose, free fatty acid, lactate, and insulin for 24 h in patients with NIDDM. *Diabetes.* 37, 1020-1024.

Rider MH, Bertrand L, Vertommen D, Michels PA, Rousseau GG, & Hue L (2004). 6-phosphofructo-2-kinase/fructose-2,6-bisphosphatase: head-to-head with a bifunctional enzyme that controls glycolysis. *Biochem. J.* 381, 561-579.

Scheepers A, Joost HG, & Schurmann A (2004). The glucose transporter families SGLT and GLUT: molecular basis of normal and aberrant function. *JPEN J. Parenter Enteral. Nutr.* 28, 364-371.

Shepherd PR (2005). Mechanisms regulating phosphoinositide 3-kinase signalling in insulin-sensitive tissues. *Acta Physiol. Scand.* 183, 3-12.

Shulman GI, Rothman DL, Jue T, Stein P, DeFronzo RA, & Shulman RG (1990). Quantification of muscle glycogen synthesis in normal subjects and subjects with noninsulin-dependent diabetes by 13C nuclear magnetic resonance spectroscopy. *New Engl. J. Med.* 322, 223-228.

Silva AP, Alves GG, Araujo AH, & Sola-Penna M (2004). Effects of insulin and actin on phosphofructokinase activity and cellular distribution in skeletal muscle. *An. Acad. Bras. Cienc.* 76, 541-548.

Simoneau JA & Kelley DE (1997). Altered glycolytic and oxidative capacities of skeletal muscle contribute to insulin resistance in NIDDM. *J. Appl. Physiol.* 83, 166-171.

Simonsen L, Stefl B, Christensen NJ, & Bulow J (1999). Thermogenic response to adrenaline during restricted blood flow in the forearm. *Acta Physiol. Scand.* 166, 31-38.

Spriet LL & Watt MJ (2003). Regulatory mechanisms in the interaction between carbohydrate and lipid oxidation during exercise. *Acta Physiol. Scand.* 178, 443-452.

Taylor R, Magnusson I, Rothman DL, Cline GW, Caumo A, Cobelli C, & Shulman GI (1996). Direct assessment of liver glycogen storage by 13C nuclear magnetic resonance spectroscopy and regulation of glucose homeostasis after a mixed meal in normal subjects. *J. Clin. Invest.* 97, 126-132.

van Schaftingen E, Hue L, & Hers HG (1980). Fructose 2,6-bisphosphate, the probably structure of the glucose- and glucagon-sensitive stimulator of phosphofructokinase. *Biochem. J.* 192, 897-901.

Wilson JE (2003). Isozymes of mammalian hexokinase: structure, subcellular localization and metabolic function. *J. Exp. Biol.* 206, 2049-2057.

Winder WW & Duan C (1992). Control of Fructose 2,6-Diphosphate in Muscle of Exercising Fasted Rats. *Am. J. Physiol.* 262, E919-E924.

Woerle HJ, Meyer C, Dostou JM, Gosmanov NR, Islam N, Popa E, Wittlin SD, Welle SL, & Gerich JE (2003). Pathways for glucose disposal after meal ingestion in humans. *Am. J. Physiol. Endocrinol. Metab.* 284, E716-E725.

Woerle HJ, Szoke E, Meyer C, Dostou JM, Wittlin SD, Gosmanov NR, Welle SL, & Gerich JE (2006). Mechanisms for abnormal postprandial glucose metabolism in type 2 diabetes. *Am. J. Physiol. Endocrinol. Metab.* 290, E67-E77.

Wu C, Kang JE, Peng LJ, Li H, Khan SA, Hillard CJ, Okar DA, & Lange AJ (2005). Enhancing hepatic glycolysis reduces obesity: differential effects on lipogenesis depend on site of glycolytic modulation. *Cell Metab.* 2, 131-140.

Yamada K & Noguchi T (1999a). Nutrient and hormonal regulation of pyruvate kinase gene expression. *Biochem. J.* 337 (Pt 1), 1-11.

Yamada K & Noguchi T (1999b). Regulation of pyruvate kinase M gene expression. *Biochem. Biophys. Res. Commun.* 256, 257-262.

In: Glycolysis: Regulation, Processes and Diseases
Editor: Paul N. Lithaw
ISBN: 978-1-60741-103-1
© 2009 Nova Science Publishers, Inc.

Chapter V

Glycolysis and the Lung

GS Maritz*
Department of Medical Biosciences, University of the Western Cape,
7535 Bellville, South Africa

Abstract

The lung is an organ with respiratory and non-respiratory functions. As such it plays a critical role in maintaining homeostasis in the body. Various cell types occur which play a role in maintaining lung structure and function. Glucose is a major energy substrate and also plays a central role in lung development. Certain cells in the lung, for example the type I pneumocytes depends largely on glycolysis for energy. Most of the glucose used by the lung is converted to lactate. The flux of glucose through the glycolytic pathway is controlled.

Apart from its role in energy metabolism, glycolysis also plays and important role in apoptosis in the lung. In addition to playing an important role in the flow of glucose through the glycolytic pathway, evidence suggests that glyceraldehyde-3-phosphate dehydrogenase, also plays a role in induction of apoptosis. In addition it also serves as an intracellular sensor for oxidative stress that may play an important early role in the cascade of reactions leading to apoptosis.

Glycolysis is also necessary for normal aging of lung cells and thus the maintaenance of lung structure and function. It has been shown that suppression of glycolysis induces premature aging in lung. This adversely affects maintenance of lung structure and function and increased susceptibility to respiratory diseases.

A number of studies showed that maternal nicotine exposure during gestation and lactation resulted in an irreversible inhibition of glycolysis. The site of action appears to be at the phosphofructokinase level. It is proposed that this inhibition is due to a change in the program that controls glucose flux though glycolysis by oxidant effects of nicotine. It is also suggested that the permanent inhibition may probably result in premature aging of the lungs of the offspring that was exposed to nicotine via the placenta and mother's

* GS Maritz: Tel: +27 21 959 2186; Fax: +27 21 959 3125; E-mail: gmaritz@uwc.ac.za

milk. This means that using nicotine replacement therapy to quit smoking during gestation and lactation is not advisable.

A. Introduction

The lung is an organ of extreme metabolic complexity. It is concerned with external gas exchange, acts as a filter for smaller particles, maintains a stable level of circulating leucocytes (Heinemann and Fishman, 1969), functions as a blood reservoir (Fishman, 1966), and is a site of intrathoracic reflexes (Dawes and Comroe, 1954).

The lung is of major importance in combating respiratory infections (Laurenzi et al, 1964). The macrophages are not only phagocytic but have bacteriolytic properties, while the bronchial secretions contain immunoglobulins (Keimowitz, 1964).

The lung is a rich source of co-factors that either promote or inhibit blood co-agulation (Heinemann and Fishman, 1969). It is a source of thromboplastin which converts prothrombin to thrombin; of an activator to convert plasminogen to plasmin; of heparin synthesized by the mast cells (Noga et al, 1999); of histamine and slow-reacting substances; and of lipoprotein lipase (Robinson and French, 1960).

The lung is also involved with the inactivation of seretonin, bradykinin, and the removal of prostaglandins from the circulation (Ferreira and vane, 1967). It activates the rennin-angiotensin system by converting angiotensin 1 to angiotensin II (Nasr and Heinemann, 1965). It furthermore participates in the *de novo* synthesis of fatty acids (Nasr and Heinemann, 1965) proteins, and of phospholipids (Yeager and Massaro, 1972; Maniscalco et al, 1978).

The above shows that the lung is metabolically a very active organ and plays a crucial role in maintaining homeostasis in the body. To fulfill its function as a gas-exchanger and in homeostasis it must have an adequate energy supply. The oxygen and glucose consumed by the lung is used is therefore, not only for its own basal metabolic needs, but also for the various non-respiratory functions mentioned above. The importance of glycolysis in the non-respiratory functions of the lung is indeed illustrated by the fact that inhibition of glycolysis in the lung result in a reduced uptake of 5-hydroxytryptamine by the lung (Steinberg et al, 1975).

Due to the heterogeneous cellular composition of the lung, the glucose and oxygen uptake by the different cellular components is still largely unknown, but it appears that the alveolar macrophages, mast cells and type II pneumocytes might be responsible for most of the oxygen uptake. The alveolar type I cells, that makes up almost 94% of the alveolar surface area (Engehardt, 2002; Naimark, 1977), have very few mitochondria and is dependent on glycolysis for energy (Massaro et al, 1975). This is reflected in the high rate of lactate production by the intact lung.

B. Intermediary Metabolism of the Lung

The lung is a metabolically active organ that is engaged in secretion, clearance and other maintenance functions that require energy and substrates for biosynthesis. These metabolic requirements of the lung are met in part by the uptake and catabolism of glucose where glucose is the major energy substrate of the lung (Fisher, 1984). Gluconeogenesis does not play a role in lung glucose supply in healthy individuals but alanine incorporation in glucose is elevated in lungs of patients with lung cancer and who are losing weight (Leij-Halfwerk et al, 2000). This implies that lungs do have the potential for gluconeogenesis. The glycogen stores in the adult lung are very limited and thus not a very good source of glucose to maintain glycolysis. The lung therefore depends on external glucose for its demands (Fisher, 1984). During severe caloric restriction glucose utilization by the lung is reduced and lipids used instead (Gregorio et al, 1976). It is therefore conceivable that the type I cells will under these conditions not be able to derive enough energy for its survival and replacement by the type II cells. This may contribute to starvation induced emphysema (Karlinsky et al, 1986)

Apart from glucose the adult lung, like most other organs, can also oxidize fatty acids, amino acids, lactate and glycerol (Fisher, 1984). Under normal physiological conditions the rate of glucose oxidation is highest in the lung in comparison to the oxidation of the other above mentioned substrates. (Kerr et al, 1979). Glucose, on entering the lung is mainly metabolized via the glycolytic pathway. About 40 to 50% of the glucose that enters the pathway exits as lactate (figure 1).

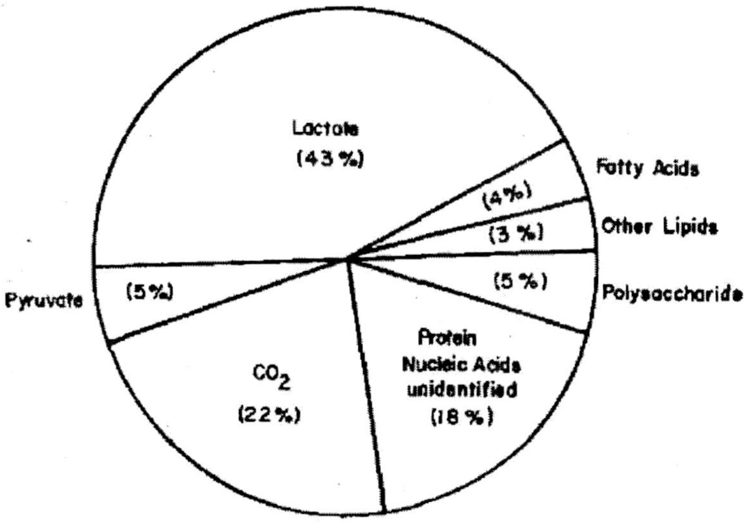

Figure 1. Recovery of carbon atoms derived from glucose metabolism during lung perfusion studies with 5.5 nM (UC^{14}) glucose in a Krebs bicarbonate medium, pH 7.4 containing 3% bovine serum albumin. The data are represented as a percentage of total recovery (Fisher, 1984).

Despite the high partial pressure of oxygen, the rate of lactate production remains consistently high, accounting for almost 10% of the total body lactate production. Thus, the rate of lactate produced is an important marker for the activity of the glycolytic pathway in the lung (Kerr et al, 1979). It is therefore plausible that persistent suppression of the glucose flux through the glycolytic pathway (figure 2) will adversely impact on the maintenance of lung structure (Peterson et al, 1984; Kerr et al, 1979) and its non-respiratory functions. Since the lungs are exposed to foreign substances in the air and blood, such as oxidants, many of these substances my severely compromise the glycolytic pathway, and thus lung structure and function. This may render the lungs more susceptible to respiratory diseases such as emphysema and even premature aging (Kondoh et al, 2005 and 2007). Since the non-respiratory functions may also be adversely affected, suppression of glycolysis in the lung may have an adverse effect on homeostasis in the entire body.

The catabolism of glucose also generate CO_2, glycrol-3-phosphate, pyruvate, and ribose which can be utilized as intermediates for the synthesis of lipids. The catabolism of glucose also yields NADPH and NADH which is essential for reductive biosynthesis and detoxification reactions (Fisher, 1984).

C. Carbohydrate Metabolism and Lung Development

1. Control of the Glycolytic Flux

Control of glycolysis can be divided into a long term control involving, for example, hormones and short term control involving regulation of enzyme activity by intermediate substrates of glucose metabolism. Two important regulators are the cellular ATP content and the cellular NAD^+ content. In lung tissue, like in other tissue a depletion of ATP and NAD^+ result in an increase in the flux of glucose through this pathway (Bassett et al, 1976). Glycolysis in the lung responds to the redox state independently of the energy status of the cell. Lactate and pyruvate, products of glycolysis, are also inhibitory when it accumulates in lung cells (Bassett et al, 1976). The regulation of glycolysis by substrates and by the energy status or redox state of the lung cell is qualitatively similar to those in other tissues (Fisher, 1984).

Studies showed that insulin control glycolysis. Glucose utilization and lactate production increase in presence of insulin (Kerr et al, 1979; Stubbs et al, 1977). This is supported by findings of Fricke et al (1979) who showed that glucose utilization in lungs of diabetic rats are less than in non-diabetic rats. More recently studies showed that stimulation of N^+-K^+-ATPase by hyperinsulinemia is associated with an increase in lactate production in skeletal muscles (Novel-Chate et al, 2001). This is consistent with studies that illustrated that aerobic glycolysis is coupled with Na^+-K^+-ATPase activity in isolated cells (Ramlal et al,1996). Na^+-K^+-ATPase also occur in the Type I and –II pneumocytes where it plays a critical role in keeping the alveoli "dry" to ensure normal gas exchange (Ridge et al, 2003; Wendt et al, 1998). It is therefore likely that the Na^+-K^+-ATPase in these cells are also linked to glycolysis to ensure optimal gas exchange. This flux of glucose through glycolysis is critical for the non-respiratory as well as the respiratory functions of the lung.

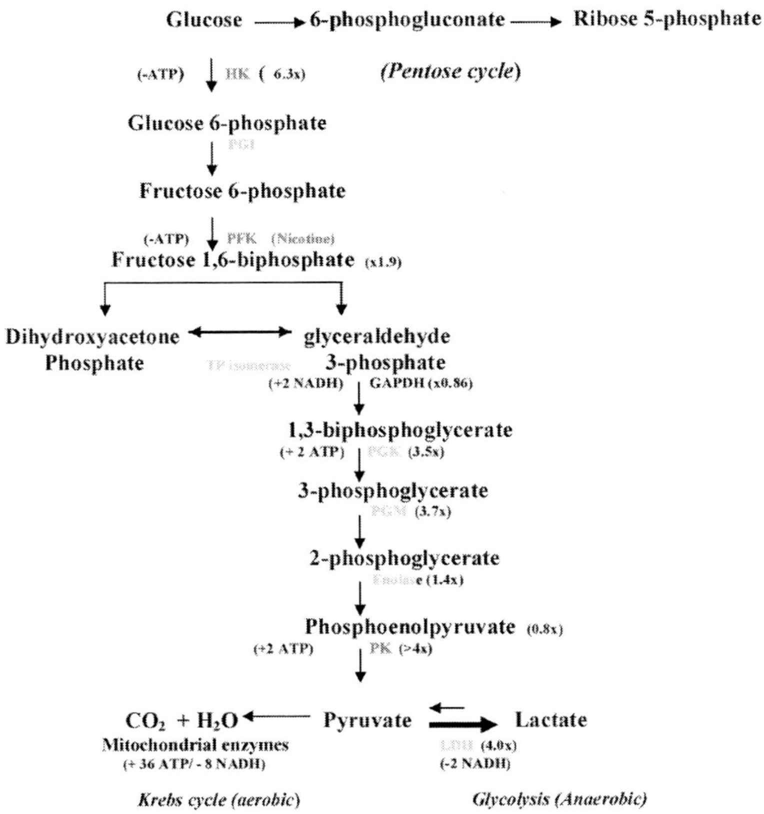

Nicotine= site of action of nicotine.

Figure 2. Alterations of the glycolytic pathway in senescent fibroblasts. The flux of glucose through the various metabolic intermediates is illustrated. Depending on the pathway, glucose can be used by the fibroblast and other cells for energy production via glycolysis, glycolysis and the Krebs cycle, and for synthesis of cellular constituents, such as the nucleotides by the pentose phosphate pathway. Glycolytic enzymes (Green and red) catalyzing each conversion are indicated. Changes in the activity of selected glycolytic enzymes and the intracellular concentration of selected metabolites, which were observed in senescent cells, are indicated by numbers. HK, hexokinase; PGI, phosphogluco isomerase; PFK, phosphofructokinase; GAPDH, glyceraldehydes-3-phosphate dehydrogenase; PGK, phosphoglycerate kinase; PGM, phosphoglycerate mutase; LDH, lactate dehydrogenase. The enzymes indicated in red are major sites of control of the flux of glucose through the glycolytic pathway.

2. Glycolysis and Lung Development

Glucose uptake and metabolism are essential for the proliferation and survival of cells, and may be enhanced in actively proliferating cell systems such as embryonic tissue. Glucose is considered to be an essential source of energy in lung tissue (O'Neill and Tierney, 1974) and is necessary for the functional development of the lung (Gilden et al, 1977; Maniscalco et al, 1978; Bourbon and Jost, 1982). Glucose is also the main source of α-glycerophosphate for surfactant synthesis in the adult lung while in fetal lung, loss of cellular glycogen from alveolar type II cells just before birth is associated with increased surfactant synthesis (Salisbury-Murphy et al, 1966). During the alveolar phase of lung development, which occurs from around week 36 of gestation in humans (Post and Copland, 2002), lung tissue is more dependent on glycogen as an energy substrate than adult lung. This is illustrated by the fact that during fasting the activity of phosphorylase of adult lung tissue decreases to conserve glycogen while the activity of phosphorylase in fetal and neonatal lung increases, thereby increasing the utilization of the lung glycogen stores. This means that the control of glycogen metabolism during the alveolar phase of lung development is different from that of adult animals (Maritz, 1988).

Although glucose and glycogen are the primary energy substrates of adult and developing lung, fatty acids are also important. For example, during fasting, when blood fatty acid levels are elevated, fatty acids replace glucose as primary energy substrate. Under these circumstances glucose is conserved by the lung for α-glycerophosphate synthesis and eventual surfactant formation by the type II pneumocytes (Rhoades, 1975).

Maternal nicotine exposure during gestation and lactation result in sustained suppression of glycolysis and glycogenolysis (figures 3 A and B) in lung tissue of the rat fetus and neonate (Maritz, 1986; Maritz, 1987). The lower glycogenolytic activity is due to a lower phosphorylase activity in the lungs of the nicotine exposed offspring (Maritz, 1986). The ratio of inactive to active phosphorylase of lung tissue of nicotine exposed offspring is the same as for animals that were not exposed to nicotine via the placenta and mother's milk. However, the tissue levels of both the phosphorylase fractions are lower than in the lungs of the control animals which implied that the total phosphorylase content of the lungs of the nicotine exposed animals was lower than that of the control animals. This means that maternal nicotine exposure suppressed the synthesis of phosphorylase in the lungs of the offspring. This might be due to changes in the epigenetic control mechanisms. It also implies that maternal nicotine exposure had no direct inhibitory effect on the activity of the phosphorylase in the lungs of the offspring. The lower rate of glycogen breakdown in the lungs of the animals that were exposed to nicotine via the placenta and mother's milk was rather due to a persistent decrease in the levels of the enzyme available to catalyze glycogenolysis than by an inhibition thereof. The implication is that the developing fetal and neonatal lungs of these animals are more dependent on exogenous glucose for utilization via the hexose monophosphate shunt than from glucose derived from the lung glycogen stores (Maritz, 1986).

The uptake of exogenous glucose is carried out by glucose transporters. Glucose transporter isoforms 1 (Glut 1) and 4 (Glut 4) are not present in adult lung, but are present in developing lung. Glut 1 is indeed expressed in type II pneumocytes and fibroblasts of the developing lung (Simmons et al, 1992). Over-expression of these Glut isoforms can enhance glucose uptake into fetal lungs to support active cell proliferation, which is a common characteristic of developing lung epithelium (Iba et al, 1999). It was indeed shown that expression of Glut 1 rapidly declines after birth (Simmons et al, 1991). This probably reflects a lower demand for glucose since the rate of cell proliferation decrease as the lungs reach maturity.

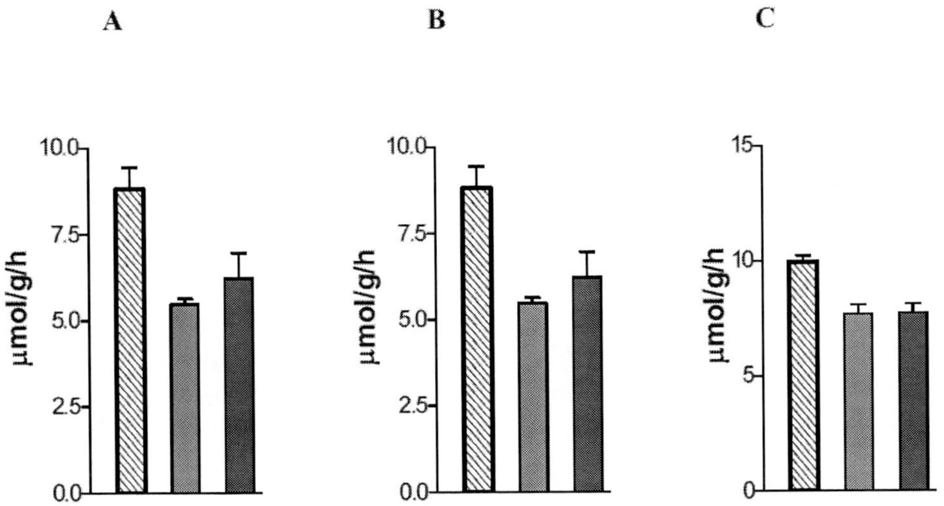

Figure 3. The influence of maternal nicotine exposure during gestation and lactation on A) glucose and B) glycogen utilization and C) lactate production by the lungs of the offspring. Withdrawal was 4 weeks after weaning of the rat pups on postnatal day 21. (Hatched bars = Control; Grey bars = Nicotine; Black bars = Withdrawal).

The decrease in the flux of glucose through the glycolytic pathway of lungs of nicotine exposed rat pups is, however, not due to a compromised glucose transporter system because the total glucose turnover of the lung tissue of rats that were exposed to nicotine via the placenta and mother's milk is higher than in the lungs of those animals that were not exposed to nicotine. The higher glucose flux is actually due to a faster utilization of glucose via the hexose monophosphate shunt (Maritz, 1983). After nicotine withdrawal the flux of glucose through the glycolytic pathway remained suppressed to the same degree than while exposed to nicotine (figure 3 A). The flux of glucose through the hexose monophosphate pathway returns to normal (Maritz, 1987). This means that phosphorylation of glucose by hexokinase is not affected by maternal nicotine exposure, nor is it affecting the hexose monophoshate pathway. It was indeed shown that maternal nicotine exposure during gestation and lactation had no effect on the expression and activity of hexokinase in lung tissue of the offspring (Maritz, 1993; Gamieldien, 2005)

Hexokinase 1 is the dominant isoenzyme in the lungs of the control as well as the nicotine-exposed rat pups. This implies that the metabolic status of the lungs of the control as well as the nicotine-exposed rats are oxidative (Hexokinase & nicotine) Apart from the role of Glut 1 and hexokinase in making glucose available for utilization by the lung cells, the regulation of the glucose carbon entry into pathways of energy production is the primary physiological role of 6-phosphofructo-1-kinase (PFK) (Mhasker and Dunaway, 1995; 1996). Three distinct subunit types have been demonstrated in rabbits (Foe and Kemp, 1984), rats (Dunaway and Kasten, 1985a; 1985b; 1987), and humans (Vora *et al.*, 1980; Khan *et al.*, 1979; Meienhofer *et al.*, 1979). PFK isozymes are the M-type in muscle and the L-type which is the major type found in the liver and the C-type which predominates in the brain and testes (Mhasker and Dunaway, 1995; 1996). The M-type predominates in organs depending on glycolysis whereas the L-type predominates in organs with active gluconeogenesis. The C-type is found in rapidly replicating cells that are largely dependent on aerobic metabolism. (Vora, 1982; 1983). Although these subunits have basically the same catalytic properties, they differ in their sensitivity to ATP inhibition and the other effectors, for example, muscle PFK (PFK-M) is less sensitive to fructose 2,6-biphosphate than liver PFK (PFK-L) (Kasten and Dunaway, 1993). This implies that control of glycolysis in various tissues will depend on the isoform that is dominant in that tissue. A change in the isoform during development reflects a change in the metabolic need of that tissue to meet the changing demand of the tissue as it develops, and thus to ensure that the development of the tissue is according to the program that directs tissue growth and development from the embryonic to the adult phase. This also applies to the developing lung and any change in the PFK isoform due to a change in the program that controls the changes of PFK as a function of growth and development will have an impact on the normal pattern of lung growth and development of the respiratory system. Recent studies showed that all three the PFK isoforms are expressed at similar levels just after gestation. However, around the critical phase of alveolarisation PFK-M, the isoform associated with glycolytic activity and PFK-C, the isoform associated with aerobic glycolysis, are expressed at similar levels. Both these isoforms are expressed at markedly higher levels than PFK-L. This implies that glycolysis becomes more important as the lungs mature while the importance of gluconeogenesis decreased (Gamieldien, 2005).

The impact of maternal nicotine exposure during gestation and lactation on the mRNA expression of PFK isoforms in developing neonatal rat lung is most obvious in the expression of PFK-M and PFK-L. The higher level of PFK-M mRNA expression suggests a higher PFK-M activity in the lungs of the nicotine exposed rat pups compared to that of the control lungs. This means that the glucose flux through the glycolytic pathway of these lungs should be higher or equal to that in the lungs of the control rats. However, contrary to this, the flux of glucose through this pathway is irreversibly suppressed (Maritz and Burger, 1992). This suggests that the PFK-M isoenzyme activity, and in all probability the total PFK activity is not higher than in lungs of control animals despite the higher mRNA expression. In a study by Kordom et al (2003) it was indeed found that the total PFK activity of rats exposed to nicotine via the placenta and mother's milk was lower than in lungs of control animals. Therefore, the need for the tissue to over-express this isoform probably stems from the inhibitory effect that maternal nicotine exposure has on the glycolytic pathway. Since glycolysis in lung tissue of nicotine exposed rats is irreversibly inhibited, it is plausible that the activity of PFK-M will be permanently lower in lung tissue of rat pups exposed to nicotine during gestation and lactation, especially since it was suggested that the lower PFK activity can be attributed to a conformational change in the structure of the enzyme (Kordom et al, 2003). Since expression of PFK-C and –L of the lung tissue of the nicotine exposed animals are the same as that of PFK-C and –L of control lung tissue, it is likely that only PFK-M was affected by maternal nicotine exposure during gestation and lactation. This also implies that the long-term effect of maternal nicotine exposure on total PFK activity can be attributed to changes in the PFK-M isoenzyme activity.

Furthermore, Dewar *et al.* (2002) showed that nicotine in liver results in a significant increase in ATP synthesis. They attribute this to the surplus of ADP as a result of reduced production of pyruvate and lactate caused by the inhibition of glycolysis. Maritz and Burger (1992) also demonstrated a marked increase in the ATP content of lung tissue of rat pups that were exposed to nicotine during gestation and lactation. As a consequence the activities of PFK-L and –C will be lower. The higher ATP content that is maintained in the lungs of the nicotine-exposed lungs will also reduce the activity of PFK-M despite a higher resistance to ATP inhibition compared to the other isoforms (Kasten and Dunaway, 1993). Although the level of the reduction might be at a lower level than that for the other isoforms, the total PFK activity will be reduced resulting in a slower flux of glucose through glycolysis.

Since the half-life of nicotine is only 90 -120 minutes (Benowitz et al, 1982) it is unlikely that there would be any nicotine in the lungs of the offspring after weaning on postnatal day 21. This means that the inhibition of glycolysis in the lungs of the offspring as a consequence of maternal nicotine exposure during gestation and lactation will be irreversible. As a consequence the over expression of PFK mRNA in nicotine-exposed lungs will also be irreversible. It is not clear whether the changes in the expression of the PFK mRNA and PFK activity contribute to the reported deterioration of the lung parenchyma in the long-term (Maritz, 2002). It is known though that a decrease in the glucose flux through glycolysis result in quicker aging of cells and tissue (Kondoh et al, 2007). This implies that the irreversible inhibition of glycolysis in the lungs of the nicotine exposed offspring will result in faster aging of the lungs of the offspring. It can be expected that age dependent deterioration, such as senile emphysema, will be faster, and age related changes in lung structure will appear at a younger age. Studies by Maritz and Windvogel (2003) indeed supported this in that maternal nicotine exposure resulted in a faster aging of the lung of the offspring which rendered it more susceptible to disease and compromised function.

In addition to inducing premature aging in lung cells, inhibition of the flux of glucose through the glycolytic pathway limits proliferation of pulmonary microvascular endothelial cells (Parra-Bonilla et al, 2008). Since the development of the pulmonary microvascular system is also important in normal alveolar formation (Burri, 2006), it is likely that suppression glucose flux through the glycolytic pathway will have an adverse effect on the reserve capacity of the lung as a gas-exchanger.

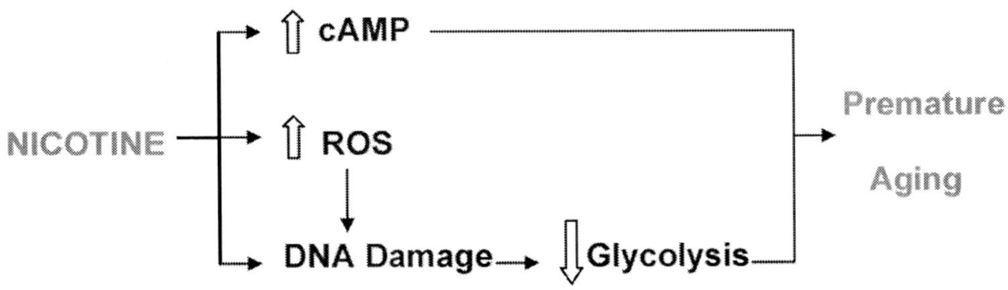

Figure 4. Diagram to illustrate the metabolic changes that induces premature aging in of the lung parenchyma of rats exposed to nicotine via the placenta and mother's milk.

In addition to the reduced flux of glucose through the glycolytic pathway (Maritz, 1987), AMP also accumulates in the lungs of the nicotine exposed rat pups and the AMP content of the lungs of the nicotine exposed offspring increased even after nicotine withdrawal (Maritz and Burger, 1992). Both the persistent reduced glycolytic activity and high levels of AMP are associated with premature onset of cell senescence (Kondoh et al, 2005, Zwerschke et al,

2003). This is supported by studies that showed that enhancement of glycolysis bypasses cellular senescence (Kondoh et al, 2007). It can therefore be expected that maternal nicotine exposure during gestation and lactation induce premature aging of the lungs of the offspring by irreversible suppression of glycolysis and the persistent high levels of AMP in the lungs of these animals (figure 4). It is plausible that the aging effect of cigarette smoke and lung fibroblasts (Nyunoya et al, 2006) are partially due to the nicotine present in tobacco smoke.

Since placental perfusion is not affected by maternal smoking and nicotine intake (Bainbridge and Smith, 2006), and since the body weight of rat pups exposed to nicotine via the placenta and mother's milk was not affected by it (Maritz and Windvogel, 2003a), it is unlikely that an inadequate blood and nutrient supply to the developing fetus induced the above mentioned changes. It is therefore likely that the metabolic changes, such as high levels of AMP and a permanently reduce glycolytic pathway in the lungs of the nicotine exposed rats induced cellular senescence which again, over time, resulted in structural changes that also resembles premature aging of the lungs. It is plausible that the lower anti-oxidant capacity of these lungs made it more susceptible to damage to the DNA and thus the "program" that control lung growth, development and aging. A consequence of cellular senescence may be a slower self-renewing of the lung cells, thus causing impaired regeneration of lung tissue. Cellular senescence may also cause disrupted tissue structure through the release of degradative enzymes (Chen et al, 2007).

3. Glycolysis and Aging

Senescent cells not only loose their ability to divide and to respond to mitogenic stimuli, but also display alterations in morphology and metabolic profile (Bird et al, 2003). This phenotype can be induced by oxidative stress (Balin et al, 2002). It is therefore plausible that factors that induce premature aging of lung fibroblasts will also affect not only growth and development of the lung, but also adversely affect the maintenance of the lung structure and function since the fibroblasts provide part of the structural support and matrix that is important for its integrity (Absher, 1995).

The fact that the lifespan of many species can be extended through caloric restriction, suggests a critical role for alterations of carbohydrate metabolism in the control of regulatory processes that influence cell proliferation and survival (Lane et al, 2001). Studies by Lin et al (2000) established a new concept according to which changes in carbohydrate metabolism, and in particular the regulation of glycolytic energy production, contribute to the control and regulation of cell proliferation and survival.

Experiments with human diploid fibroblasts (HDF) showed a drastic deregulation of carbohydrate metabolism in senescent cells. This is characterized by an imbalance of glycolytic enzyme activities and the failure to maintain ATP levels. This resulted in up-regulation of adenylate kinase and of the levels of AMP, which is known to act as a growth-suppressive signal that induces premature senescence (Zwerschke et al, 2003).

Zwerschke et al (2003) showed that about 90% of the consumed glucose in young HDF is converted to lactate. Thus, these cells displayed aerobic glycolysis, characterized by a high rate of lactate production from glucose in the presence of oxygen. The remaining 10% is used for ATP production and for synthetic processes. It is interesting to note that most of the glucose consumed by lung tissue is also changed to lactate (Peterson et al, 1984, Kerr et al, 1979). The lactate is produced by the type I alveolar epithelial cells and in all likelihood the interstitial fibroblasts. Other cells such as the type II alveolar epithelial cells and alveolar macrophages have a very active Krebs cycle and respiratory chain which implies that most of the glucose consumed by these cells will be converted to CO_2 and H_2O and energy (Kerr et al, 1979). Since the Type II cells continuously produce surfactant (Botas et al, 1998), it is important for these cells to have a continuous supply of precursors and energy via the glycolytic pathway and Krebs cycle and respiratory chain.

Studies with human fibroblasts showed that age-dependent changes in the metabolism include an increased rate for the conversion of glucose to alanine by senescent fibroblasts along with lactate. These studies suggest that glycolysis is up-regulated in senescent human fibroblasts. This is supported by observations by Zwerscke et al (2003), namely that, in senescent HDF the specific activity of hexokinase, phosphoglycerate kinase, phosphoglycerate mutase, pyruvate kinase and lactate dehydrogenase was increased. The up-regulation of these enzymes can be expected to considerably increase the flux of glucose through the glycolytic pathway. However, in contrast to this assumption, there was a drastic reduction in the flux of glucose through glycolysis of the senescent fibroblasts. This implies that other adjustments must have occurred to this pathway that prevent the expected increase in glucose flux through glycolysis in senescent cells. This is probably due to the fact that GAPDH and enolase was not upregulated. It is therefore plausible that the failure of senescent cells to maintain aerobic glycolysis is due to their inability to co-ordinately regulate the activity of the glycolytic enzymes (figure 2). Consequently the fructose 1,6-biphosphate increased strikingly in these HDF. This can be ascribed to the increase in HK and the decreased activity of GAPDH (Zwerschke, et al, 2003). The upregulation of the ATP consuming part of the glycolytic pathway and the decrease in the generation of ATP below GAPDH resulted in a reduced ATP content in these cells. It is likely that this impairment in the generation of ATP via glycolysis is due to the capturing of the phosphate in phosphometabolites such as fructose 1,6-biphosphate. It implies that the drop in ATP content will go with an increase in the AMP content of these cells.

It was indeed shown that the AMP levels increased drastically in senescent HDF (Lee et al., 2001). This increase is in all likelihood due to the upregulation of adenylate kinase (AK) in these cells. This up-regulation of AK was also observed in aging rats (Lee et al, 2001). They also demonstrated that the expression of genes involved in energy generating pathways in duodenum of aging rats, specifically cytochrome c oxidase, ATP synthase, and sodium-potasium ATPase was down-regulated. Studies by Ethier et al (1989) also showed an increase in adenosine release from cultured human lung fibroblasts from aged donors, which is likely due to an enhanced breakdown of ATP in these fibroblasts. The higher levels of AMP serve as a strong anti-proliferative signal on the cells and aging of the fibroblasts. It is therefore conceivable that changes in metabolic control which will result in a down-regulation of glycolysis and of and increase in AMP will result in premature aging of the lung fibroblasts and other lung cells. This suggestion is supported by Müller et al (2006) who showed that lung fibroblasts from patients with emphysema display markers of aging.

4. Glycolysis and Apoptosis

Glucose metabolism plays a critical role in the protection of a variety of cell types against oxidant-induced cell death (Moley and Mueckler, 2000). The hexokinases play an important role in both the uptake and utilization of glucose by catalyzing the first committed step of glucose metabolism by catalyzing phosphorilation of the glucose molecule. They also initiate the major pathways of glucose utilization and are ideally positioned to influence the metabolic flux through both glycolysis and the pentose phosphate pathway (Bryson et al, 2002). Increased expression of hexokinase is also associated with decreased susceptibility of renal epithelial cells to oxidant-induced cell death and is compatible with the hypothesis that the hexokinases play an important role in the anti-apoptotic effects of growth factors (Bryson et al, 2002). They demonstrated that ectopic expression of hexokinase is associated with improved cell morphology, decreased cell detachment, decreased apoptosis, and reduced cytolysis in H_2O_2-stressed cells. Since maternal nicotine exposure had no long-term effect on hexokinase in the lungs of rat pups that were exposed to nicotine via the placenta and mother's milk, and since it had no impact on the flux of glucose through the pentose phosphate shunt, it is unlikely that it will have an adverse effect on the protection of the lungs via the hexose monophosphate shunt. In some pathological conditions other factors induce apoptosis by decreasing glucose transport and in turn trigger the glucose deprivation apoptosis cascade (Moley and Mueckler, 2000). On the other hand, overexpression of GLUT1 prevented an increase in JNK and induction of apoptosis (Moley and Mueckler, 2000). However, since glucose flux through the pentose phosphate shunt of the lungs of the rats that were exposed to nicotine via the placenta and mother's milk was not suppressed, it is clear that glucose transport into the lung cells as well as phosphorylation by hexokinase, was not affected. It is therefore unlikely that apoptosis was affected by inadequate glucose entry into the metabolic pathways of the lungs of these animals. However, since glycolysis is irreversibly suppressed (Maritz, 1987) in the lungs of the nicotine exposed offspring, it can be expected that protection of the lungs of the nicotine-exposed rats against apoptosis by glycolysis will be adversely affected. The importance of glycolysis is demonstrated by the fact that suppression of glycolysis with inhibitors 2-deoxyglucose and iodoacetate result in a reversal of the protective effect of glucose (Long et al, 1997; Fujio et al, 1997)

Glyceraldehyde-3-phosphate (GAPDH) is a glycolytic enzyme with a key role in energy metabolism (Chuang et al, 2004). Since the lung, and especially the type I pneumocytes depends on glycolysis for energy (Massaro et al, 1975), this enzyme plays and important role in the survival of these cells. Apart from its role in energy metabolism, GAPDH also display other functions independent of its function in glycolysis. These include regulation of the cytoskeleton (Huitorel and Pantolani, 1985; Fuchtbauer et al, 1986), membrane fusion and transport (Glaser and Gross, 1995; Robins et al, 1995; Tisdale, 2001), and more. There is mounting evidence that GAPDH is an integral part of apoptosis. It is suggested that p53 directly induce GAPDH expression and in this way apoptosis (Chuang et al, 2004; Tarze, 2007). This suggests that there is a possible link between glycolysis and apoptosis (Malhotra and Brosius, 1999). Bcl-2 blocks the pro-apoptotic role of GAPDH. This implies that Bcl-2 participates in the regulation of GAPDH and its role in protecting cells against apoptosis (Maruyama et al, 2002).

It has been thought that GAPDH serves as an intracellular sensor of oxidative stress and may play an early important role in the cascade leading to apoptosis (Hara et al., 2001). This result in an inhibition of GAPDH (Brune and Mohr, 2001) and glucose flux continues through the pentose phosphate shunt which generates NADPH used by glutathione reductase to recycle oxidized glutathione to its reducing form. It also uncouples glucose metabolism from the production of ATP and oxidative intermediates (Buchczyk et al, 2000; Buchczyk et al, 2003). This means that under oxidative conditions GAPDH act as a switch to redirect glucose metabolism from the glycolytic pathway to the pentose shunt.

Maternal nicotine exposure during gestation and lactation, in other words during all the phases of lung development, resulted in an inhibition of GAPDH activity (Maritz, 1997). Apoptosis plays an important role in the thinning of the alveolar walls of the developing lung. It can thus be assumed that the lower GAPDH activity in the lungs of the rats that were exposed to nicotine during gestation and lactation will result in a slower apoptosis and thus a slower thinning of the alveolar walls. This supports the suggestion that the slower thinning of the alveolar walls of rat pups that were exposed to nicotine via the placenta and mother's milk was due to suppression of apoptosis (Heusch and Maneckjee, 1998).

In addition to suppression of apoptosis, nicotine also induce lipid peroxidation (Ashakumary, Vijayammal, 1996) and reduces the antioxidant capacity of the lungs of the offspring by reducing the vitamin C content of the lungs of adult rats (Maritz 1993) as well as superoxide dismutase activity (Windvogel, et al, 2008). Since GAPDH acts as a sensor for increased oxidant activity with a resultant flux of glucose through the pentose phosphate shunt to regenerate NADPH, it is plausible that the increased flux of glucose via the pentose shunt while the animals were exposed to nicotine can partly be attributed to this protective function of GAPDH. After nicotine withdrawal GAPDH and the flux of glucose through the pentose shunt return to normal (Maritz, 1997). This can probably be ascribed to a decrease in the level of lipid peroxidation since nicotine is not present in the lungs of these animals anymore.

Apart from GAPDH, other enzymes of the glycolytic pathway also participate in apoptosis. This is illustrated by the observation that the expression of C-myc, an oncogene, increase in most human cancers, including lung carcinoma. Expression of c-myc is followed by upregulation of LDH-1. Overexpression of LDH-1 in fibroblasts cause apoptosis during glucose deprivation. This suggests that LDH-1 links c-myc to glucose dependent apoptosis in lung. It is suggested that constitutive generation of NAD^+ and lactate by LDH-1 and the decline in NADH due to inhibition of glycolysis alters the redox state of the cells, triggering the apoptotic pathway (Moley and Mueckler, 2000). It is therefore likely that the permanent decrease in glucose flux through glycolysis in lungs of animals that were exposed to nicotine during gestation and lactation will have an enhanced apoptosis. This interference with the factors that control apoptosis will in the long term result in alveolar damage such as emphysema.

Conclusion

Glycolysis is crucial for lung growth and development especially since the Type I cells depends on glycolysis for survival. It is also important for fibroblast function and as precursor for synthesis of surfactant by the Type II pneumocytes. Apart from the importance of the flux of glucose through the glycolytic pathway, individual enzymes plays an important role in maintaining cell structure and survival. Any suppression of glycolysis or of individual enzymes will adversely affect lung growth, development and maintenance of respiratory health in the long term. It is therefore, essential to maintain a life style that will prevent especially the developing lung from exposure to substances that will suppress glycolysis. This include use of nicotine replacement therapy during pregnancy and lactation.

References

Absher, M. (1995) Fibroblasts: In lung cell biology. Lung Biology in health and disease. Vol. 41. Ed by Massaro D, Marcel Dekker. Inc. New York, Basel; p401-439.

Ashakumary, L; Vijayammal, PL. Effect of Nicotine on Antioxidant Defence Mechanisms in Rats Fed a High-Fat Diet. *Pharmacology.* 1996; 52: 153-158.

Bainbridge, SA; Smith, GN. The effect of nicotine on in vitro placental perfusion pressure. *Can. J. Physiol Pharmacol.* 2006; 84: 953-957.

Balin, AK; Fisher, AJ; Anzeloni, M; Leong, I; Allen, RG. (2002) Effects of establishing cell cultures and cell culture conditions on the proliferative life spn of human fibroblasts isolated from different tissues and donors of different ages. *Exp. Cell Res.* 274: 275-287.

Bassett, DJP; Fisher, AB. Stimulation of rat lung metabolism with 2,4-dinitiphenol and phenazine ethosulfate. *Am. J. Physiol.* 1976; 321: 898-902.

Benowitz, NL; Kuyt, F; Jacob III, P. Circadian blood nicotine concentrations during cigarette smoking. *Clin. Pharmac. Ther.* 1982; 32: 758-764.

Bird, J; Ostler, EL; Faragher RGCan we say that senescent cells cause ageing? *Exp. Gerontol.* . 2003; 38: 1319-1326.

Botas, C; Poulain, F; Akiyama, J; Brown, C; Allen, L; Goerke, J; Clements, J; Carlson, E; Gillespie, AM; Epstein, C; Hawgood, S. Altered surfactant homeostasis and alveolar type II cell morphology in mice lacking surfactant protein D. *Proc. Nat. Acad. Sci.* (USA) 1998; 95: 11869-11874.

Bourbon, J; Jost, A. Control of glycogen metabolism in the developing fetal lung. *Pediatr. Res.* 1982; 16: 50-56.

Brune, B; Mohr, S. Protein thiol modification of glyceraldehydes-3-phosphate dehydrogenase and caspase-3 by nitric oxide. *Curr. Protein Pept. Sci.* 2001; 2: 61-72.

Buchczyk, DP; Briviba, K; Harti, FU; Sies, H. Response to peroxynitrite in yeast: glyceraldehde-3-phosphate dehydrogenase (GAPDH) as a sensitive intracellular target for nitration and enhancement of chaperone expression and ubiquination. *Biol. Chem.* 2000; 381: 121-126.

Buchczyk, DP; Grune, T; Sies, H; Klotz, LO. Modifications of glyceraldehydes-3-phosphate dehydrogenase induced by increasing concentrations of peroxynitrite: early recognition by 20S proteosome. *Biol. Chem.* 2003; 237-241.

Burri, PH. Structural Aspects of Postnatal Lung Development - Alveolar Formation and Growth. *Neonatology.* 2006; 89: 313-322.

Chuang, D-M; Hough, C; Senatorov, VV. Glyceraldehyde-3-phosphate dehydrogenase, apoptosis and Neurodegenarative diseases. *Annu. Rev. Pharmacol. Toxicol.* 2004; 45: 269-290.

Dawes, GS; Comroe, JH. Chemoreflexes from the heart and lungs. *Physiol. Rev.* 1954; 34: 167-201.

Dewar, BJ; Bradford, BU; Thurman, RG. Nicotine increases hepatic oxygen uptake in the isolated perfused rat liver by inhibiting glycolysis. *J. Pharmacol. Exp. Ther.* 2002; 301:930-937.

Dunaway, G.A; Kasten, TP. Characterization of the rat heart 6-phosphofructo-1-kinase isoenzymes. *J. Mol. Cell Cardiol.* 1985a; 17:947-957.

Dunaway, GA; Kasten, TP. Nature of the rat brain 6-phosphofructo-1-kinase isoenzymes. *J. Biol. Chem.* 1985b; 260:4180-4185.

Dunaway, GA; Kasten, TP. Nature of the subunits of the 6-phosphofructo-1-kinase isoenzymes from rat tissue. *Biochem. J.* 1987; 251:677-683.

Engelhardt, JF. The lung as a metabolic factory for gene therapy. *J. Clin. Invest.* 2002; 110: 429-432.

Ethier, MF; Hickler, RB; Dobson Jr, JG. Aging increases adenosine and inosine release by human fibroblast cultures. *Mech. Ageing Dev.* 1989; 50: 159-168.

Ferreira, SH; Vane, JRC. The disappearance of bradykinin and clodoisin in the circulation. *Nature.* 1967; 216: 868-869.

Fishman, AP. Trends in Pharmacological Sciences; Pulmonary hypertension. *Circulation.* 1966; 33: 835

Fisher, AB. Intermediary metabolism of the lung. *Envirin. Health Perspect.* 1984; 55: 149-158.

Foe, LG; Kemp, RG. Isozyme composition and phosphorylation of brain phosphofructokinase. *Arch. Biochem. Biophys.* 1984; 228:503-510.

Fricke, RF; Longmore, WJ. Effect of insulin and diabetes on 2-deoxy-D-glucose uptake by the isolated pefused rat lung. *J. Biol. Chem.* 1979; 254: 5092-5098.

Fuctbauer, A; Jockusch, BM; Leberer, E; Pette, D. Actin-severing activity copurifies with phosphofructokinase. *Proc. Natl. Acad. Sci. USA*. 1986; 83: 9502-9506.

Fujio, Y; Kunisada, K; Hirota, H; Yamauchi-Takihara, K; Kishimoto. T. Signals through gp-130 upregulated bcl-x expression via STAT1-binding cis-element in cardiac myocytes. *J. Clin. Invest.* 1997; 99 2898-2905.

Gaddum, JH; Hebb, CO; Silver, A; Swan, AAD. 5-hydroxytryptamine; pharmacological action and destruction in perfused lungs. *Quart. J. Exper. Physiol.* 1953; 38: 255-262.

Gamieldien, k; Maritz, GS. Influence of maternal nicotine exposure during gestation and lactation on lactate dehydrogenase isoenzyme profile and transcript levels in developing neonatal rat lung. *Pathophysiology.* 2008: 15: 1-8.

Gamieldien, K. The influence of maternal nicotine exposure on selected glycolytic and cytochrome P450 enzymes in developing neonatal rat lung. PhD thesis, University of the Western Cape, South Africa. 2005; p74-76.

Gilden, C; Sevanian, A; Tierney, DF. Regulation of fetal lung phosphatidylcholine synthesis by cortisol: role of glycogen and glucose. *Pediatr. Res.* 1977;11: 845-848.

Glaser, PE; Gross, RW. Rapid plasma phenylethanolamine-selective bilayers catalyzed by an isoform of glyceraldehydes-3-phosphate dehydrogenase-discrimination between glycolytic and fusogenic roles of individual isoforms. *Biochemistry.* 1995; 34: 12193-12203.

Gregorio, CA; Gail, DB; Massaro, D. Influence of fasting on lung oxygen consumption and respiratory quotient. *Am. J. Physol.* 1976; 230: 291-294.

Hara, M; Agarawal, N; Luo, X; Fujii, K; Snyder, SH; Sawa, A. Oxidative modification of GAPDH and neuronal cell death; a trigger to initiate a death cascade via Siah and N-coR. *Annu. Meet. Soc. Neurosci.* Abstr. No.19.11.

Heinemann, HO; Fishman, AP. Non-respiratory functions of the mammalian lung. *Physiol. Rev*. 1969; 49:1-7

Heusch, WL ; Maneckjee, R. Signalling pathways involved in nicotine regulation of apoptosis of human lung cancer cells. *Carcinogenesis.* 1998; 19: 551-556

Huitorel, P; Pantaloni, D. Bundling of microtubules by glyceraldehydes-3-phosphate dehydrogenase and its modulation by ATP. *Eur. J. Biochem.* 1985; 150: 265-269.

Iba, MM; Pak, YW; Thomas, PF. Dose dependent upregulation of rat pulmonary regulation of rat pulmonary, renal, and hepatic cytochrome P450 (CYP) 1A expression by nicotine feeding. *Drug Metab. Dispos.* 1999; 2: 977-82.

Kahn, A; Meienhofer, MC; Cottreau, D; LaGrange, JL; Dreyfuss, JC. Phosphofructokinase (PFK) isoenzymes in man. I. Studies in adult human tissue. *Human Genet.* 1979; 48:93-108.

Karlinsky, JB; Goldstein, R H; Ojserkis, B; Snider, G L. Lung mechanics and connective tissue levels in starvation-induced emphysema in hamsters. *Am. J. Physiol. Regul. Integr. Comp. Physiol* .1986;251: R282-R288.

Kasten, TP; Dunaway,GA. Fructose 2,6-bisphosphate: changes during neonatal maturation and ageing of rat and postnatal role in regulation of glucose utilization. *Mech. Ageing and Devel.* 1993; 68:37-45.

Keimowitz,RI. Immunoglobulins in normal human tracheo-bronchial washings. *J. Lab. Clin. Med.* 1964; 63: 54-59

Kerr, JS; Baker, NJ; Bassett, JP; Fisher, AB Effect of perfusate glucose concentration on rat lung glycolysis. *Am. J. Physiol.* 1979; 236: E229-E233.

Kondoh, H; Lleonart, ME; Bernad, D; Gil, J. (2007) Protection from oxidative stress by enhanced glycolysis; a possible mechanism of cellular immortalization. *Histol. Histopathol.* 22: 85-90.

Kondoh, H; Lleonart, ME; Gil, J; Wang, J; Degan, P; Peters, G; Martinez, D; Carnero, A; Beach,D. Glycolytic enzymes can modulate cellular life span. *Cancer Res*. 2005; 65: 177-185.

Kordom, C; Maritz, GS; de Kock ,M. Maternal nicotine exposure during pregnancy and lactation: I. Effect on glycolysis in the lungs of offspring. *Exp. Lung Res.* 2003; 29:79-89.

Lane, MA; Black, A; Handy, A; Tilmont, EM; Ingram, DK; Roth, GS. Caloric restriction in primates. *Ann. NY Acad. Sci.* 2001; 928: 287-295.

Laurenzi, GA; Berman, L; First, M; Kass, EH. A quantitative study of the deposition and clearance of bacteria in the murine lung. *Clin. Invest.* 1964; 43: 759-768.

Lee, HM; Greeley Jr, GH; Englander, EW. Age-associated changes in gene expression patterns in the duodenum and colon of rats. *Mech. Ageing Dev.* 2001; 122: 355-371.

Leij-Halfwerk, S; Dagnelie, PC; van den Berg, JWO; Wattimena, JDL; Hordijk-Luijk, C; Wilson, JHP. Weight loss and elevated gluconeogenesis from alanine in lung cancer patients. *Am. J. Clin. Nutr.* 2000; 71: 583-589.

Long, X; Boluyt, MO; Hipolito, ML; Lundberg, MS; Zheng, JS; O'Neill, L; Cirieli, C; Lakatta, EG; Crow, MT. P53 and the hypoxia-induced apoptosis of cultured neonatal rat cardiac myocytes. *J. Clin. Invest.* 1997; 99: 2635-2643.

Malhotra, R; Brosius, FC 3rd. Glucose uptake and glycolysis reduce hypoxia-induced apoptosis in cultured neonatal rat cardiac myocytes. *J. Biol. Chem.* 1999; 274:12567-12575.

Maniscalco, WM; Wilson, CM; Gross, I. Development of glycogen and phospholipid metabolism in fetal and newborn rat lung. *Biochem. Biophys. Acta.* 1978; 530: 333-346.

Maritz, GS; Burger, B. The influence of maternal nicotine exposure on neonatal lung carbohydrate metabolism. *Cell Biol. Int. Rep.* 1992; 16:1229-1236.

Maritz, GS. Maternal nicotine exposure during gestation and lactation of rats induce microscopic emphysema in the offspring. *Exp. Lung Res.* 2002; 28: 391-403.

Maritz, GS; Windvogel S. Chronic maternal nicotine exposure during gestation and lactation and the development of the lung parenchyma in the offspring: Response to nicotine withdrawal. *Pathophysiology.* 2003; 10: 69-75.

Maritz, GS. Die invloed van nikotien op die intermediêre metabolisme van longweefsel (The influence of nicotine on the intermediary carbohydrate metabolism of lung tissue) PhD. Thesis, University of Stellenbosch, South Africa. 1983;

Maritz, GS. Pre- and postnatal carbohydrate metabolism of rat lung tissue. *Arch. Toxicol.* 1986; 59: 89-93.

Maritz, GS. Maternal nicotine exposure and carbohydrate metabolism of fetal and neonatal lung tissue: response to nicotine withdrawal. *Respiration.* 1987; 51: 232-240.

Maritz, GS. Lung glycogen metabolism in suckling rats: A comparative study. *Biol. Neonate.* 1988; 54: 100-106.

Maritz, GS. Influence of maternal nicotine exposure on key enzymes of glucose metabolism in lung tissue of the offspring. *Pathophysiology.* 1997; 4: 135-141.

Maritz, GS. The influence of maternal nicotine exposure on neonatal lung metabolism. Protective effect of ascorbic acid. *Cell Biol. Internat.* 1993; 17: 579-584.

Massaro, GD; Gail, DB; Massaro, D. Lung oxygen consumption and mitochondria of alveolar epithelial and endothelial cells. *J. Appl. Physiol.* 1975; 38: 588- 592.

Maruyama, W; Oya-Ito, T; Shamoto-Nagai, M; Osawa, T; Naoi, M. Glyceraldehyde-3-phosphate dehydrogenase is translocated into nuclei through Golgi apparatus during apoptosis induced by 6-hydroxydopamine in human dopaminergic SH-SY5Y cells. *Neurosci. Lett.* 2002; 3211: 29-32.

Meienhofer, MC; LaGrange, D; Cottreau, D; Lenoir, G; Dreyfus, JC.. (1979) Phosphofructokinase in human cells. *Blood.* 1979; 54:389-400.

Mhaskar, Y; Dunaway, GA; Alteration of PFK subunit protein, synthesis, and mRNA during neonatal brain development. *Dev. Brain Res.* 1995; 85:54-57.

Mhaskar, Y; Dunaway, GA. Alteration of 6-phosphofructo-1-kinase subunit protein, synthesis rates, and mRNA during rat neonatal development. *Mech. Ageing Develop.* 1996; 86:161-172.

Moley, KH; Mueckler, MM. Glucose transport and apoptosis. *Apoptosis.* 2000; 5: 99-105

Müller, K-C; Welker, L; Paasch, K; Feindt, B; Erpenbeck, VJ; Hohlfield, JM; Krug, N; Nakashima, M; Branscheid, D; Magnussen, H; Jörres, RA; Holz, O. (2006) Lung fibro*blasts from patients with emphysema show markers of senescence in vitro.* Resp. Res. 2006; 7: 32-36.

Naimark, A. Non-ventilatory functions of the lung. *Am. Rev. Respir. Dis.* 1977; 115: 93-98.

Nasr, K; Heinemann, HO. Lipid synthesis in rat lung tissue. *Am. J. Physiol.* 1965; 208: 18-121.

Ng, KKF; Vane, JR. Conversion of angiotensin 1 to angiotensin II. *Nature.* 1965; 216: 762-766.

Noga, O; Brunnée, T; Schäper, C; Kunkel, G. Heparin, Derived from the Mast Cells of Human Lungs Is Responsible for the Generation of Kinins in Allergic Reactions due to the Activation of the Contact System. *Int. Arch. Allergy Immunol.* 1999;120:310-316 (DOI: 10.1159/000024284)

Novel-Chate, V; Valentine, R; Chiolero, R; Schneitner, P; Leverve, X; Jequier, E; Tappy, L. Role of Na+-K+-ATPase in insulin-induced lactate release by skeletal muscle. *Am. J. Physiol. Endocrinol. Metab.* 2001; 280: E296-E300.

Nyunoya, T; Monick, MM; Klingelhutz, A; Yarovinsky, TO; Cagley, JR; Hunninghake, GW. Cigarette smoke induces cellular senescence. *Am. J. Resp. Cell and Mol. Biol.* 2006; 35: 681-688.

O'Neill, JJ; Tierney, DF. Rat lung metabolism: Glucose utilization by isolated perfused lungs and tissue slices. *Am. J. Physiol.* 1974; 226: 867-873.

Parra_Bonilla, G; Alvarez, DF; Cioffi, DL; Stevens, T. Glucose restriction inhibits glycolytic flux and limits proliferation in lung microvascular endothelial cells. *FASEB J.* 2008; 22:1178.10

Peterson, DR; Norris, KJ; Thompson, JA; A comparative study of the disposition of nicotine and its metabolites in the three inbred strains of mice. *Drug Metab. Dispos.* 1984;12: 725-731.

Post, M; Copland I. Overview of lung development. *Acta Pharmacol. Sin.* 2002; 23 Suppl: 4-7

Ramlal, T; Ewart, HS; Somwar, R; Deems, RO; Valentin, MA; Young, DA; Klip, A. Muscle subcellular localization and recruitment by insulin of glucose transporters and Na^+-K^+-ATPase subunits in transgenic mice overexpressing the GLUT4 glucose transporter. *Diabetes.* 1996; 45: 1516-1523.

Ridge, KM; Olivera, WG; Saldias, F; Azzam, Z; Horowitz, S; Rutschman, DH;,Dumasius, V; Factor, P; Sznajder, JI. Alveolar Type I cells express the α2 Na,K-ATPase, which contributes to lung liquid clearance. *Circ. Res.* 3003; 92: 453-460.

Rhoades, RA. Net uptake of glucose, glycerol and fatty acids by the isolated perfused rat lung. *Am. J. Physiol.* 1975; 226: 144-149.

Robinson, DS; French, JE. Heparin, the clearing factor lipase, and fat transport. *Pharmacol. Rev.* 1960; 12: 241-263.

Robinson, AR; Ward, RD; Oliver, C. A mutation of glyceraldehydes-3-phosphate dehydrogenase alters endocytosis in CHO cells. *J. Cell Biol.* 1995; 130: 1093-1104.

Salisbury-Murphy, S; Rubinstein, D; Beck, JC. Lipid metabolism in lung slices. *Am. J. Physiol.* 1966; 211: 988-992.

Simmons, RA; Gounis, AS; Bangalore, SE; Ogata, ES. Intrauterine growth retardation: fetal glucose transport is diminished in lung but spared in brain. *Pediatr. Res.* 1992; 31: 59-63.

Simmons, RA; Gounis, AS; Bangalore, SA; Ogata, ES. Regulation of glucose transport in type II pneumocytes of fetal and neonatal rats. *Pediatr. Res.* 1991; 29: 52A (Abstr).

Steinberg, H; Bassett, DJP; Fisher, AB. Metabolic requirements for hydroxytryptamine uptake by the isolated lung. *Chest.* 1975; 67: 59-60.

Stubbs, WA; Morgan, I; Lloyd, B; Alberti, KBM. The effect of insulin on lung metabolism in the rat. *Clin. Endocrinol.* 1977; 7: 181-184.

Tarze, A; Deniaud, M; Le Bras, E; Maillier, D; Molle, N; Larochette, N; Zamzami, G; Kroemer, G; Brenner, C. (2007). "GAPDH, a novel regulator of the pro-apoptotic mitochondrial membrane permeabilization". *Oncogene.* 26 (18): 2606–2620

Tisdale, EJ. Glyceraldehyde-3-phosphate dehydrogenase is required for vesicular transport in the early secretory pathway. *J. Biol. Chem.* 2001; 276: 2480-2486.

Vora, S; Seaman, C; Durham, S; Piomelli, S; Isozymes of human phosphofructokinase: identification and subunit structural characterization of a new system. *Proc. Natl. Acad. Sci. USA.* 1980; 77:62-66.

Wendt, CH; Sharma, R; Bair, R; Towle, H; Ingbar, DH. Oxidant effects on epithelial Na, K-ATPase Gene expression and promoter function. *Environ. Hlth Perspect.* 1998; 106: 1213-1217.

Windvogel, S; Maritz, G; Gamieldien, K; Gelderblom, W. Maternal nicotine exposure and neonatal lung development: Investigating the modulating effect of rooibos (asphalatus linearis) on oxidative status of rat lung.Abstract; 2008: MRC Research Day, p107.

Yeager, H; Massaro, D. Glucose metabolism and glycoprotein synthesis by lung slices. *J. Appl. Physiol.* 1972; 477-482.

Zwerschke, W; Mazurek, S; Stökl, P; Hütter, E; Eigenbrodt, E; Jansen-Dürr, P. (2003) Metabolic analysis of senescent human fibroblasts reveals a role for AMP in cellular senescence. *Biochem. J.* 376: 403-411.

In: Glycolysis: Regulation, Processes and Diseases
Editor: Paul N. Lithaw, pp. -
ISBN: 978-1-60741-103-1
© 2009 Nova Science Publishers, Inc.

Chapter VI

Transcriptional and Post-Transcriptional Regulation of Glycolysis in Microbial Cells

***Dave Siak-Wei Ow*[1,2], *Victor Vai-Tak Wong*[1] *and Andrea Camattari*[1]**

[1] Bioprocessing Technology Institute, 20 Biopolis Way, #06-01 Centros, Singapore 138668

[2] NUS Graduate School for Integrative Sciences and Engineering, 28 Medical Drive, #05-01, National University of Singapore, Singapore 117456

Abstract

Escherichia coli, Bacillus subtilis and *Saccharomyces cerevisiae* are well-characterized species which have contributed significantly to our present knowledge of central metabolism. In addition to their roles as model organisms in biology, they are also widely used as microbial cell factories for the biotechnological production of valuable products like insulin and vaccines. Glycolysis is the core pathway for carbon metabolism in these cells to provide the necessary energy and carbon backbones for product synthesis and cellular growth. Carbon fluxes through glycolysis have evolved to be under rigid regulatory control so as to coordinate catabolic fluxes with biosynthetic demands during growth. While the control of activity of glycolytic enzymes through allosteric regulation is well-understood, the regulation of glycolytic genes at the transcriptional level has begun to attract attention only recently. Additionally, a few post-transcriptional regulators were also found to regulate glycolysis at the level of mRNA stability. This communication will describe our current knowledge on glycolysis-related transcriptional/post-transcriptional factors regulating mRNA synthesis and degradation in these three representative microbial cell systems.

1. Introduction

With few exceptions, glycolysis is the main route for central carbon metabolism in nearly all organisms. Starting with glucose-6-phosphate and ending with pyruvate, the glycolytic pathway plays important dual physiological roles in catabolism and anabolism. The end product of glycolysis, pyruvate feeds into the TCA cycle, which then generates high-energy molecules and biosynthetic precursors to support physiological growth. These fundamental steps of glycolysis are generally ubiquitous in both eukaryotic and prokaryotic cells.

Escherichia coli, Bacillus subtilis and *Saccharomyces cerevisiae* are three well-characterized species representing the main microbial genres of Gram-negative bacteria, Gram-positive bacteria and Crab-positive yeast respectively. Due to their ease of genetic manipulation in the laboratory, they have become standard model organisms that have contributed extensively to our present knowledge of biochemistry and regulation of metabolic pathways. The regulation of metabolic pathways could be modulated at the level of (post)transcription, (post)translation or protein activity (allosteric regulation). Allosteric regulation of glycolysis by binding of small metabolites to their effecter sites on glycolytic enzymes is already well known (Table 1). More recently, the regulation of glycolytic genes at the transcriptional and post-transcriptional level by various specific and global regulatory proteins has begun to attract attention from the scientific community. This communication will review the current knowledge on transcriptional and post-transcriptional regulators of glycolysis in *E. coli, B. subtilis* and *S. cerevisiae.*

Table 1. Allosteric regulation of enzyme activity in glycolysis and related anaplerotic pathways [1]

Enzyme	Activator	Inhibitor
Phosphofructokinase 1 (PfkA)	ADP	PEP
Fructose-1,6-biphosphate (Fbp)		AMP
Pyruvate kinase 1 (PykF)	FBP	
Pyruvate kinase 2 (PykA)	AMP	
Phosphoenolpyruvate carboxylase (Ppc)	FBP, Acetyl-CoA	Aspartate, MAL
Phosphoenolpyruvate carboxykinase (PckA)		NADH

* Abbreviations: Acetyl-CoA, Acetyl-Coenzyme-A; FBP, Fructose-1,6-biphosphate; ADP, Adenosine-diphosphate; AMP, Adenosine-monophosphate; PEP, Phosphoenolpyruvate; MAL, Malate; NADH, Nicotinamide adenine dinucleotide (reduced).

2. Transcriptional and Post-Transcriptional Regulation of Glycolysis in *E. Coli*

2.1. Cyclic AMP Response (CRP) Protein

The entry of many carbon sources into the glycolytic pathway in *E. coli* relies on the phosphoenolpyruvate: sugar phosphotransferase system (PTS), which transport carbon sources such as glucose, fructose, lactose and mannose through a multi-step carbohydrate phosphorylation and membrane translocation process [2]. Catabolite repression is the phenomenon that allows *E. coli* to utilize glucose before less preferred carbon sources, and it directly controls the entry of carbon molecules feeding into the glycolytic pathway. Cyclic AMP response protein (CRP) is the principal and earliest-known pleiotropic transcriptional regulator of catabolite repression [3]. The CRP regulator is activated in the presence of an intercellular alarmone cyclic AMP (cAMP), which accumulates during starvation conditions or the absence of glucose. Upon binding with cAMP, the cAMP-CRP complex is able to directly regulate at least 197 genes belonging to several functional classes [4].

Currently, the cAMP-CRP complex is known to be a transcriptional activator of three glycolytic genes: *gapA, fbaA, pgk* (figure 1). Based on similarity to consensus sequence and primer extension studies, cAMP-CRP complex was first found to activate the transcription of a glycolytic gene *gapA* [5] encoding for glyceraldehyde-3-phosphate dehydrogenase (GAPDH). GAPDH plays an important role in carbon metabolism by catalyzing the reversible conversion of D-glyceraldehyde-3-phosphate into 1,3-diphosphoglycerate. Interestingly, the *E. coli gapA* product is structurally related to eukaryotic GAPDHs. Subsequently, a cAMP-CRP consensus promoter site was also found before the *fbaA* and *pgk* genes encoding for two other glycolytic enzymes, fructose-bisphosphate aldolase and phosphoglycerate kinase. Deletion of the cAMP-CRP promoter reduced phosphoglycerate kinase activity by 2.4 fold, revealing that more than half of *pgk* expression depends on transcriptional activation from the cAMP-CRP promoter site [6].

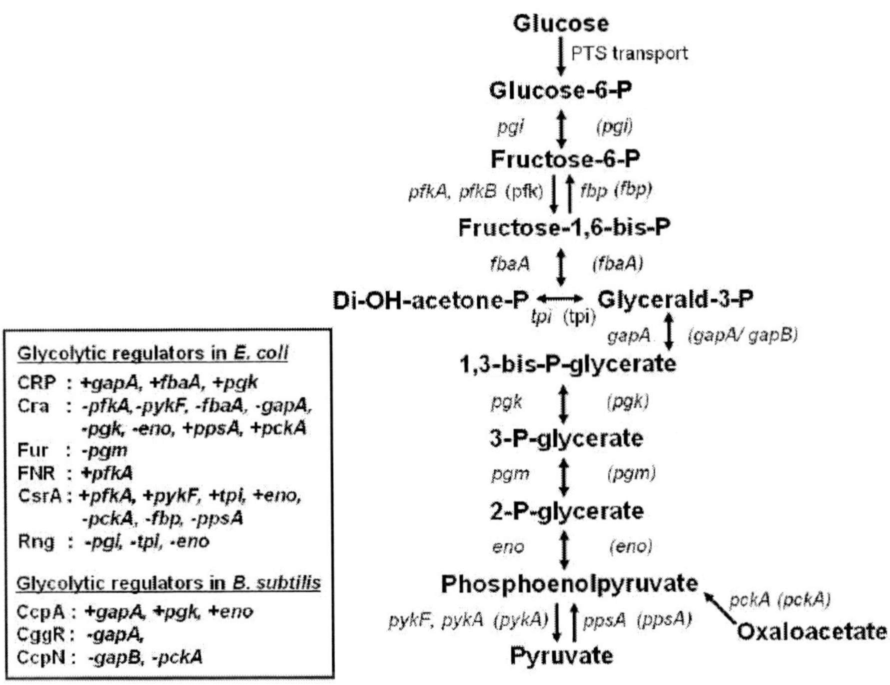

Figure 1. Transcriptional and post-transcriptional regulators of glycolysis in *Escherichia coli* and *Bacillus subtilis.* Positive regulation are indicated by a + sign while negative regulation are indicated by – sign. Genes are shown in italics. Within the glycolytic pathway on the right, *E. coli* genes are not in brackets, while the corresponding *B. subtilis* genes are in brackets.

2.2. Catabolite Repressor Activator (Cra)

The catabolite repressor activator (Cra), also known as fructose repressor (FruR), is another pleiotropic transcriptional regulator of carbon metabolism acting independently of the cAMP-CRP complex in *E. coli* [7]. Cra represses the synthesis of several enzymes involved in the utilization of carbon substrates though glycolytic pathway and activates enzymes involved in oxidative or gluconeogenic (the reverse direction from glycolysis) pathways [8]. The regulatory activity of Cra is feed-back modulated by the intercellular levels of two glycolytic intermediates: fructose-1-phosphate or fructose-1,6-diphosphate. Low concentrations of fructose-1-phosphate (micromolar levels) or fructose-1,6-diphosphate (millimolar levels) cause the tetrametric Cra protein complex to completely disassociate from its operator sites [9,10,11]. In line with its regulatory effects on the glycolysis, a knockout of the fruR gene increases glycolytic gene expression [12], enhances carbon flow though the glycolytic pathway, and inhibits carbon flow though glyconeogenesis [10].

The search for the Cra-binding consensus sequences close to the -35/-10 promoter region [9,13] led to the identification of several Cra-regulated genes related to glycolysis or glyconeogenesis (Table 2). The transcriptional activation of the *E. coli* glyconeogenetic gene *ppsA* (catalyzing the conversion of pyruvate into phosphoenolpyruvate) was first revealed from the gene expression, DNA foot-printing and gel mobility shift assays [9]. Subsequently, the *in-vitro* binding of Cra to the operator sites of the glycolytic *pykF* and the glyconeogenetic *pckA* genes were later confirmed using DNA band mobility shift assays [10,11]. The *pykF* gene encodes for the last enzyme of glycolysis (pyruvate kinase I) while *pckA* encodes for the glyconeogenetic enzyme (phosphoenolpyruvate carboxykinase) that convert oxaloacetate from TCA cycle into phosphoenolpyruvate. Regulation of *pgk* and *fbaA* by Cra were also implicated based on a putative FruR binding site at the promoter region of the *gapB-pgk-fbaA* transcript [6] and that the expression of a *gapB-lacZ* fusion was shown to be repressed by Cra [10].

While initial searches for Cra-regulated genes using consensus sequences were done manually, subsequent use of high-throughput systematic methods for the identification of other consensus sequences led to the discovery of more Cra-regulated glycolytic genes. From a systematic search of Cra-binding sequences from a collection of random synthetic sequences using CAST (cyclic amplification and selection of targets), 20 palindrome-like DNA sequences that binds Cra with high affinity were found. The examination of these DNA sequences produced a new Cra-binding consensus sequence and led to the detection and subsequent confirmation of a Cra-binding site before the *pfkA* gene coding for the first enzyme of glycolysis, phosphofructokinase I [14]. Using another systematic approach known as SELEX (systemic evolution of ligands by exponential enrichment) which uses a library of genomic DNA fragments, four other Cra- regulated targets were identified and verified [15]. Among these were two new glycolytic genes: *eno* (enolase) and *gapA* (glyceraldehyde-3-phosphate dehydrogenase gene).

Table 2. Cra-regulated genes in glycolysis and glyconeogenesis

Genes	Enzyme	Regulation	References
Glycolysis			
pfkA	Phosphofructokinase I	Repression	14
pykF	Pyruvate kinase I	Repression	10,11
gapA	Triosephosphate dehydrogenase (GAPDH-A)	Repression	15
eno	Enolase	Repression	15
pgk	Phosphoglycerate kinase	Repression	6
fbaA	Fructose-bisphosphate aldolase II	Repression	6
Glyconeogenesis			
ppsA	Phosphoenolpyruvate synthetase	Activation	9
pckA	Phosphoenolpyruvate carboxykinase	Activation	10

2.3. Ferric Uptake Regulator (Fur) and Fumarate and Nitrate Reductase Regulation (FNR) Proteins

The ferric uptake regulator (Fur) is a negative global transcriptional regulator, which employs iron as a co-repressor and represses several operons involved in iron transport and several other cellular functions [16,17]. Using a strain carrying a chromosomal reporter controlled by a Fur-regulated promoter, Fur-regulated loci exhibiting weak affinity for the Fur repressor were identified [17]. Of these, the promoter for the glycolytic gene *pgm* encoding was subsequently shown to contain Fur-binding sequences (fur boxes) and its promoter activity was iron-regulated and Fur-dependent.

The fumarate and nitrate reductase regulation (FNR) protein is an oxygen-responsive transcriptional regulator crucial for the switch from aerobic to anaerobic metabolism [18]. FNR senses cytoplasmic oxygen levels via its oxygen-sensitive Fe-S cluster in the N-terminal domain, and the presence of oxygen reacts with the Fe-S cluster and rapidly inactivates the regulatory activity of FNR. Structurally related to CRP, FNR has a broad regulon spectrum including its role as a positive regulator of anaerobic fermentative genes [19,20].

There were some evidences indicating that FNR could be a transcriptional activator of the glycolytic *pykA* gene in *E. coli* during microaerobic conditions. Although the binding of FNR to the promoter region of *pykA* has not been verified, during microaerobic chemostat growth when the headspace oxygen concentrations were between 1-5%, the inactivation of FNR was shown to reduce *pykA* expression strongly by more than three fold compared to the wildtype cells [21]. The *pykA* gene code for pyruvate kinase II, which is an isoenzyme of the Cra-regulated pyruvate kinase I (*PykF*). Both pyruvate kinase isoenzymes catalyze the final irreversible glycolytic step of converting phosphoenolpyruvate into pyruvate. Differential regulation between these two glycolytic isoenzymes has been previously demonstrated in several gene expression studies [22,23,24]. This suggests that *pykA* and *pykF* could be under the control of different transcription regulators.

2.4. Post-Transcriptional Regulation by Carbon Storage Regulator (Csr) and Rnase G

The carbon storage regulator A (CsrA) is a small 7 kDa RNA-binding protein that regulates various exponential and stationary phase processes of *E. coli* at the post-transcriptional level [25]. The apparent mechanism of post-transcriptional regulation involves the binding of CsrA to the 5'leader of the untranslated mRNA, which could either (1) inhibit protein translation and facilitates rapid degradation of the repressed mRNA [26,27,28], or conversely, (2) stabilize the steady-state mRNA levels of the activated genes [29].

In addition to the repression of several stationary phase metabolic processes like gluconeogenesis (*pckA, fbp, ppsA*) and glycogen synthesis, CsrA is also an activator of a number of glycolytic genes. From lacZ-transcriptional fusion and enzyme activity comparison of *csrA*$^-$ versus *csrA*$^+$ strains, the glycolytic genes *pfkA, pykF, tpi* and *eno* were found to be positively activated, while *pfkB* was repressed and *pykA* was not affected [30]. Interestingly, the two pairs of isoenzyme genes for phosphofructokinase (pfkA, pfkB) and pyruvate kinase (*pykF, pykA*) exhibited differential regulation by CsrA. Phosphofructokinase and pyruvate kinase catalyze the first and the last glycolytic reaction steps and these are the only two irreversible reactions in glycolysis [31]. Due to that, they are generally regarded as key regulation points for glycolysis. The differential post-transcriptional regulation by CsrA on the isoenzyme genes for phosphofructokinase and pyruvate kinase should, therefore, allow further modulation at the gene expression level to the allosteric regulation of glycolysis [30].

RNase G (Rng) is the latest proposed post-transcriptional regulator of glycolysis. It is a RNA-specific endonuclease first found to assist in the 5' maturation of 16S ribosomal RNA [32]. With the discovery of more RNA targets, RNase G is recognized to play a regulatory role as well by facilitating cellular messenger and regulatory RNA turnover [33].

The first evidence that RNase G affects glycolytic mRNA stability came from a DNA microarray study of rng mutants. Out of 4405 open reading frames, 11 mRNA transcripts which steady-state level was influenced by cellular concentration of RNase G were found [33]. These included three glycolytic genes *pgi, tpi* and *eno*, which mRNA transcript level increased in abundance following depletion of RNase G. The regulation of *eno* expression by RNase G was independently further verified in another study, whereby two-dimensional gel electrophoresis analysis on an *rng* knockout strain showed a 2-3 fold increase in the Eno protein [34]. Rifampicin chase experiments also revealed that the half-life of the *eno* transcript was 3.2 fold more in the *rng* knockout strain over the parental strain, hence affirming that the *eno* mRNA is an *in-vivo* substrate of RNase G. Recently, a combinational mutation of RNase G and Cra in an *E. coli* strain was found to increase the production of the glycolytic end product pyruvate [35]. This was attributed to the coordinated derepression of glycolytic gene expresson by RNase G and Cra.

3. Transcriptional Regulation of Glycolysis in *Bacillus Subtilis*

The genes for the glycolytic reaction sequence from fructose-6-phosphate to pyruvate have been identified in *B. subtilis* and are illustrated in figure 1. The entry of sugars into the glycolytic pathway involves sugar transport and phosphorylation and the genes involved are inducible by the specific sugars present in the medium. The transport mechanism and the regulation of catabolic operons for individual carbohydrates have been reviewed in a recent article [36]. Like *E. coli*, the entry of carbon sources into the cell is initiated by the PTS, which senses sugar availability and transports the sugar into the cell. Unlike *E. coli*, where sugar in the extracellular environment triggers catabolite repression through cAMP and CRP (reviewed earlier in this chapter), in *B. subtilis,* catabolite repression is triggered when glycolytic intermediates from glucose metabolism, such as fructose-1,6-bisphosphate (FBP), bind to the allosteric site of the ATP-dependent kinase, HPr kinase, which activates its catalytic site [37]. Upon activation, HPr kinase phosphorylates the seryl residue (Ser-46) of the HPr protein [38] as well as its regulatory paralog Crh [39]. Ser-46 phosphorylated HPr (HPr-Ser-46-P) subsequently binds to and activates the pleiotropic transcriptional regulator, CcpA (catabolite control protein A) [40]. In addition, the interaction of HPr-Ser-46-P with CcpA requires the presence of fructose 1,6-bisphosphate [40]. Thus, FBP provides a link between glycolytic activity and the CcpA transcriptional regulator.

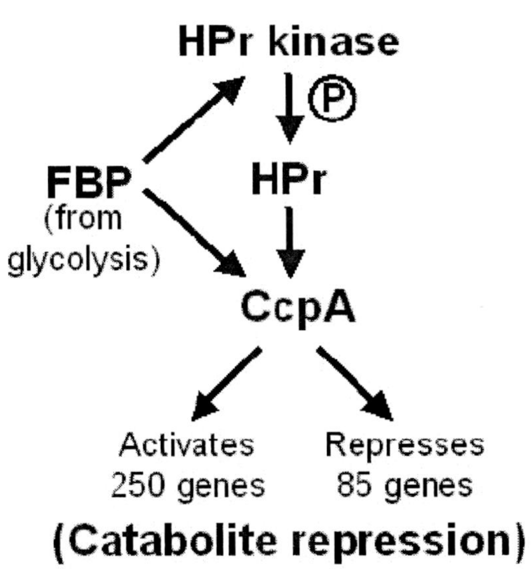

Figure 2. Activation of CcpA regulator by the glycolytic intermediate FBP. FBP activates the HPr kinase, which in turn phosphorylates the HPr protein. The phosphorylated HPr and FBP are both required to activate CcpA, which is the pleiotropic transcriptional regulator of catabolite repression.

Although the enzymes required for the interconversion of hexose phosphates and the subsequent generation of triose phosphate (Pgi, Pfk and FbaA) are constitutively expressed in *B. subtilis* [41], subsequent studies suggest that the *pfk-pykA* operon is weakly inducible by glucose [42]. An earlier study by Tobisch *et al.* [41] proposed that the *gapA* (also known as *gap*) gene (coding for glyceraldehyde-3-phosphate dehydrogenase) forms an operon with the upstream *yvbQ* gene, while the remaining genes *pgk* (phosphoglycerate kinase), *tpi* (triose phosphate isomerase), *pgm* (phosphoglycerate mutase) and *eno* (enolase) form another operon. However, subsequent studies revealed that *gapA, pgk, tpi, pgm* and *eno* transcribed as a hexacistronic operon together with *yvbQ* (renamed *cggR* for central glycolytic gene regulator). In addition, the authors showed that there are two glyceraldehyde-3-phosphate dehydrogenase genes in *B. subtilis*; *gapA* is active in glycolysis while *gapB* is active in gluconeogenesis [43]. In the following sections, we will review the current knowledge on the two key transcriptional factors regulating glycolytic enzymes in *B. subtilis*, CcpA and CggR, as well as a regulator of the gluconeogenic enzymes, CcpN.

3.1. Carbon Catabolite Protein A (CcpA)

CcpA is a member of the LacI/GalR family of transcriptional regulators [44]. Mutations in *ccpA* lead to loss of carbon catabolite repression of many catabolic genes and operons. The primary role of CcpA is to regulate genes in carbon metabolism in response to intracellular metabolite levels. Through whole-genome analysis, about 250 and 85 genes have been found to be subject to CcpA-dependent transcription repression or activation respectively [45]. Among these genes, CcpA has been implicated in the transcriptional activation of the glycolytic genes *gapA*, *pgk* and *eno* in the presence of glucose [41,45].

For most catabolic genes and operons, CcpA binds to a *cis*-acting palindromic sequence known as *cre* (catabolic responsive element) to mediate its effect [46,47]. Biochemical and structural studies have shown that CcpA binds to *cre* sequences as a complex consisting of one *cre* duplex, one CcpA dimer, and two HPr-Ser-46-P molecules [48]. The affinity of CcpA for several *cre* sites is enhanced in the presence of HPr-Ser-46-P [49]. Although *gapA* shows the putative *cis*-acting palindromic sequence known as *cre* (catabolic responsive element) [45], recent studies suggest that the effect of CcpA on the *gapA* operon may not be through direct binding in the promoter region (Ludwig et al.,2002). Instead, it has been proposed that CcpA affects the *gapA* operon through its influence on the PTS-mediated transport of sugars [50]. *ccpA* mutants show increased HPr kinase activity, which causes the HPr protein to be trapped in HPr-Ser-46-P state, thus impairing PTS sugar transport [50]. Through the modulation of the transport of PTS sugars, CcpA controls the formation of intracellular effectors (such as fructose-1,6-bisphosphate) which are cofactors for other regulators (such as CggR, described in the next section), thus indirectly causing the repression or activation of the target genes or operons.

CcpA is also involved in the regulation of TCA cycle enzymes. In contrast to glycolytic enzymes, genes in the TCA cycle, including *citH, citC, citB, odhA, sucCD* and *citG* are repressed in the presence of glucose. Tobisch *et al.* [41] showed that glucose repression of these genes is partially relieved in a *ΔccpA* mutant. In addition to regulating carbon metabolism in response to glucose, CcpA is also involved in the glucose-independent regulation of ammonium assimilation [51], stress response chaperones [41] and other genes with roles in carbon metabolism or other functions [45].

3.2. Central Glycolytic Gene Regulator (CggR)

CggR is a member of the SorC/DeoR family of transcriptional regulators and acts as a repressor of the hexacistronic *gapA* operon [43]. In the absence of glucose, CggR binds to its target DNA sequence upstream of the *gapA* gene and blocks the transcription of genes in the *gapA* operon. When glucose is present in the medium, FBP binds to a low-affinity site on CggR and releases the repressor from the DNA, allowing *gapA* expression [52,53]. Thus, the mechanism for CggR repression of *gapA* is through direct binding to the DNA sequence, which is in contrast to CcpA described above. Doan and Aymerich [52] showed that the target sequence for CggR on the *gapA* operon consist of two direct-repeats ($CGGGACN_6TGTCN_4CGGGACN_6TGTC$). CggR binds as a dimer to each direct repeat, thus forming a tetramer, and has a stronger affinity for the 3' repeat [54].

CggR has an N-terminal DNA-binding domain and a C-terminal effector-binding domain [55]. The C-terminal region of CggR regulates the DNA binding activity of this repressor in response to the binding of a phosphorylated sugar, such as FBP [56]. The effector molecule, FBP, has a bimodal effect on CggR binding to the DNA operator sequence. At micromolar concentrations, FBP binding to CggR causes a change in the conformational dynamics of the CggR-DNA complex. Upon increase to the millimolar range, FBP reduces the affinity and cooperativity of CggR binding to the full operator DNA, thus releasing the CggR repressor from the DNA [54]. The dual effect of FBP on CggR is due to the presence of two distinct sugar binding sites with different affinities for FBP [53]. FBP bound to the low affinity site acts as an inducer of transcription, while FBP bound to the high affinity site acts as a structural cofactor for the repressor [53].

The 3'end of the CggR mRNA transcript is subject to post-transcriptional processing, resulting in a short CggR transcript and a longer transcript for the pentacistronic mRNA consisting of the remaining genes (*gapA, pgk, tpi, pgm* and *eno*) in the operon [50]. Meinken *et al.* [57] demonstrated that the mRNA for the truncated *cggR* gene was quickly degraded, whereas the downstream processing products were relatively more stable. The increased stability was due to a stem-loop structure at the 5'end of the processed mRNAs. The differential stability of the segments contributes to the higher expression of GapA compared to the CggR repressor [57].

3.3. Control Catabolite Protein of Gluconeogenic Genes (CcpN)

CcpN has been recently identified as an additional regulator of catabolite repression in *B. subtilis* acting on genes encoding 2 gluconeogenic enzymes, GapB (NADP(H)-dependent glyceraldehyde-3-phoshate dehydrogenase) and PckA (PEP carboxykinase) [58]. Subsequent transcriptome analysis showed that *gapB* and *pckA* are the only protein-coding genes directly repressed by CcpN [59]. In addition, CcpN also represses the transcription of the untranslated regulatory RNA, SR1, involved in arginine catabolism [60].

CcpN is required for efficient growth under glycolytic conditions. It binds specifically to the *gapB* and *pckA* promoter regions and represses transcription in the presence of glucose or other glycolytic carbon sources [58]. The operator sites of CcpN at each of the three regulated promoters have been characterized. Each operator has two binding sites for CcpN, with one contacting more strongly than the other [61]. *ccpN* forms an operon with *yfqL* that is constitutively expressed under both glycolytic and gluconeogenic conditions [58]. As this leads to a constant occupation of all operators with CcpN, the activity of CcpN is modulated by the binding of small effector ligands. Unlike CcpA and CggR described above, the glycolytic intermediate FBP does not affect CcpN repression. Instead, ATP was found to enhance CcpN-mediated repression at the three regulated promoters, and the effect is more pronounced at a slightly acidic pH [62]. Furthermore, ADP can specifically counteract the effect of ATP to relive the repression by CcpN. Licht *et al.* [62] showed that at acidic pH, ATP significantly altered the structure of CcpN but did not affect CcpN affinity for its operators. Therefore, the authors proposed that the effect of ligand-bound CcpN on RNA polymerase may be due to a conformational switch that alters the interaction between these two proteins. Recently, Tannler *et al.* [63] demonstrated that the knockout of CcpN to deregulate expression of GapB and PckA had a positive effect on riboflavin production in an industrial strain of *B. subtilis*.

4. Transcriptional and Post-Transcriptional Regulation of Glycolysis in *Saccharomyces Cerevisiae*

Saccharomyces cerevisiae represents one of the most studied eukaryotic organisms to date. Its ability to grow on different carbon sources reflects its sophisticated carbon source regulation. In this section, the main transcriptional factors affecting glycolytic gene expression will be described; beside that, recent evidences describing the global regulation of such a central process will be taken in consideration. Some recent articles, however, require us to address the regulation of glycolysis from a possible point of debate.

With the upsurge of global expression analysis, several attempts of defining the transcriptional regulation of glycolytic genes took place [64]. Although yeast glycolysis presents several convenient features as a model of study (as a central and unbranched pathway), its complete understanding, integrated in the cellular dynamics is far from understood [65]. In particular, the quantitative correlation between the three dogmatic elements representing the cornerstones of modern biology (DNA, mRNA and proteins), revealed several biases. Although proper cultivation techniques were applied to measure glycolytic activities, perfect correlation between mRNA levels, enzymatic activity and *in-vivo* metabolic fluxes could not be determined. Conversely, it has been shown that much of the regulation of glycolytic enzymes in *S. cerevisiae* occurs post-transcriptionally during steady-state glucose-limited chemostat culture [66]. Although focusing on the mere transcriptional event that occur in the cell might miss the overall process integration, there are still several elegant articles showing that the transcriptional regulation of glycolytic enzymes [64] retain its validity. In describing the factors involved in yeast glycolytic regulation, only factors affecting transcriptional regulation will be taken into account.

4.1. GCR1 and RAP1 as the Core Regulatory Element

GCR1 is the first glycolysis-related transcriptional factor identified and characterized [67,68,69]. It encodes for a protein of the apparent molecular weight of 94 KDa, with the a low codon bias index, typical of regulatory genes in *S. cerevisiae* [68]. Its role is crucial for the transcription of glycolytic genes: a *gcr1* mutant present a 5- to 50-fold reduction in the mRNA levels for genes involved in central carbon metabolism. The GCR1 binding site lies within the Upstream Activation Sequence (UAS) in all known glycolytic genes. Even if other multifunctional factors like ABF1 or REB1 sometimes contribute to UAS activity [70], the consensus motifs for these factors can be found upstream of genes involved in other cellular functions. On the other hand, the peculiarity of GCR1 lies in its strong preference for UAS of glycolytic genes.

gcr1 mutants are characterized by small-sized colonies and a severe reduction in the transcription of most glycolytic enzymes; the binding site for GCR1 has been described by a typical CTTCC motif, named CT box [71]. The CT box has been typically found upstream of glycolytic genes, always in association with the consensus for RAP1, a more general transcriptional factor for glycolytic and ribosomal genes. The close link between the consensus motifs of GCR1 and RAP1 suggested a cooperative interaction among the two proteins; indeed, a strong synergism takes place among GCR1 and RAP1, reporting a more structural role for the latter, being RAP1 able to stretch the DNA structure to allow GCR1 to accommodate in the resulting pocket [72,73]. Interestingly, three binding sites for Gcr1p have been identified in the -200/ -100 region of the GCR1 promoter itself, suggesting the possibility of an autoregulation mechanism to ensure a constant presence of Gcr1p inside the nucleus [74].

4.2. Additional Elements: GRC2 and SGC1

As on-going evidences were clarifying the interaction between GCR1 and RAP1, additional genetic analysis and gain-of-functions mutation experiments contributed to enrich the picture and accumulated further information on transcriptional regulation of glycolysis. From that, two more factors GCR2 and SGC1 were found to be relevant in transcriptional control of several other glycolytic genes.

GCR2 was initially identified from mutation studies impairing the expression of several glycolytic genes, and it is distinct from known GCR1 mutants [75]. After cloning and characterization, it has been shown that GCR2 does not interact directly with DNA and is required for the specialized activation of the RAP1/GCR1 complex, likely through a GCR2-mediated phosphorylation event [76] leading to a subsequent conformational change of GCR1 [77]. Leucine zipper domains present on both GCR1 and GCR2 are an important feature of these transcriptional factors. According to the mechanism proposed by Stephen Deminoff and George Santangelo, either GCR1 or GCR2 are required to form homodimers before they are able to activate ribosomal or glycolytic gene expression. In the specific case of glycolytic genes specifically, GCR2 acts as a "specialized" activator when both CT box and RAP1 binding site are present. This is not required for ribosomal genes expression.

Like GCR2, SGC1 (acronym of "suppressor of *gcr*"*1*) was also identified from mutation experiments since its dominant, gain-of-function mutation suppressed the functional defects of a *gcr1* null mutant [78]. Its sequence revealed a helix-loop-helix with substantial similarity with other DNA-binding domains; its function might be achieved through the previous activation of GCR1-RAP1, followed by recruitment of SGC1 in place to support the transcription event [79].

4.3. MIG1-Regulated Glucose Repression and the Metabolic/Non-Metabolic Role of HXK2

As previously mentioned, *S. cerevisiae* is able to grow on a large variety of substrates. The phenomenon of lowering the enzyme synthesis rate for alternative carbon catabolism in presence of a preferred carbon source is called catabolite repression. As glucose is most widely known to present such a phenomenon, it is often referred to as glucose repression. The protein expressed by the gene MIG1 plays a pivotal role in regulating glucose repression and other cellular functions, as extensively reviewed by Klien and coworkers [80]. MIG1 had been cloned and characterized as a DNA repressor element for glucose repression [81]; in turn, the MIG1 promoter has been reported to be autoregulated, and DNAse I protection experiments showed a binding for Mig1p on its own promoter [82].

The glucose repression cascade, with MIG1 as the DNA-binding protein, is mainly constituted by two kinease: a serine/threonine protein kinase (Snf1p) and a sensor kinase (hexokinase, Hxk2p). Kinases play a central role in effecting glucose repression, as either sensors or effectors. The Hxk2p hexokinase, in particular, besides catalyzing the conversion of glucose into glucose-6-phosphate, acts also as a trigger for glucose repression cascade, activating the DNA repressor element for glucose repression, Mig1p. Glucose transporters, and in particular the gene products encoded by HXT2, HXT4 and SNF3, showed a strong MIG1-regulated glucose repression [83], together with the genes encoding for transporters of less conventional carbohydrates (e.g. melibiose, sucrose, galactose and mannose).

As already mentioned, the Hxk2p hexokinase not only catalyzes the whole glucose assimilation, but also the glucose repression signaling pathway. Being a crucial enzyme in a very robust pathway, the proposal of a "non-metabolic" role for HXK2 has been surprising [84]. It has been recently observed that when Hxk2p is prevented from entering the nucleus, glucose repression could not take place [85,86]. This opens up a whole scenario regarding the molecular interactions in the nucleus for the transcriptional control of glucose repression. The identity of the interacting partner for Hxk2p is still under debate. Although present evidences had suggested an interaction between Hxk2p and the protein kinase Snf1p, other targets have also been examined. Noteworthy, a direct interaction between Hxk2p and Mig1p has been proposed [87].

5. Conclusions

As the main route for carbon breakdown and energy generation in the majority of organisms, the glycolytic pathway needs to be precisely regulated to synchronize biosynthetic precursor and energy generation with various physiological demands. Although allosteric regulation plays a dominant role in the metabolic regulation of glycolysis, transcriptional and post-transcriptional regulation is now recognized to play parallel roles regulating the expression of metabolic enzymes in microbial cells. The frequent observations of glycolytic gene expression changes in response to physiological changes affirm the involvement of significant (post)transcriptional regulation of glycolysis.

The last two decades has seen the discovery of several pleiotropic (global) regulators affecting glycolysis and related pathways. This review of (post)transcriptional regulation of glycolysis in *E. coli, B. subtilis* and *S. cerevisiae* revealed interesting resemblances as well as some differences. By far, *E. coli* is the most well understood organism, hence more is known about the transcriptional and post-transcription regulation of glycolysis in *E. coli* than any other organism. On top of allosteric regulation by glycolytic metabolites and ADP/AMP, the isoenzymes (phosphofructokinase and pyruvate kinase) catalyzing the first and the last reaction steps of glycolysis are also subjected to differential (post)transcriptional regulation by at least two pleiotropic regulators, Cra and CsrA. For *B. subtilis*, the specific transcriptional regulators for phosphofructokinase and pyruvate kinase have not been found yet. As exemplified from the different routes adopted to control catabolite repression in the three organisms, *B. subtilis* and *S. cerevisiae* could have evolved diverse means to regulate the activity of these two key glycolytic enzymes, possibly also involving undiscovered interactions by other (post)transcriptional regulators.

Another common theme utilized in microbial cells is the modification of regulatory activity of several glycolytic regulators by the binding of pathway metabolites (like FBP) or small molecules including cAMP, O_2, iron and ATP. This would allow the fine-tuning of glycolytic gene expression to different physiological or environmental conditions. Further systematic investigation looking into the quantitative correlation between mRNA, proteins and metabolic fluxes under controlled physiological conditions should reveal more regulators and the dynamic interplay between different modes leading to the overall integration of key regulatory processes.

References

[1] Keseler IM, Collado-Vides J, Gama-Castro S, Ingraham J, Paley S, Paulsen IT, Peralta-Gil M, Karp PD. (2005). EcoCyc: a comprehensive database resource for *Escherichia coli*. *Nucleic Acids Res.* 33:D334-D337.

[2] Postma PW, Lengeler JW, Jacobson GR. (1993). Phosphoenolpyruvate:carbohydrate phosphotransferase systems of bacteria. *Microbiological Rev.* 57:543-594.

[3] Busby AK, Buc H, Garges S, Adhya S. (1993). Transcriptional regulation by cAMP and its receptor protein. *Annual Review of Biochemistry.* 62:749-797.

[4] Martı´nez-Antonio A. Collado-Videsy J. (2003). Identifying global regulators in transcriptional regulatory networks in bacteria. *Curr.Opin.Microbiol.* 6:482–489.

[5] Charpentier B, Branlant C. (1994). The *Escherichia coli* gapA gene is transcribed by the vegetative RNA polymerase holoenzyme E sigma 70 and by the heat shock RNA polymerase E sigma 32. *J. Bacteriol.* 176:830–839.

[6] Charpentier B, Bardey V, Robas N, Branlant C. (1998). The EIIGlc protein is involved in glucose-mediated activation of *Escherichia coli gapA* and *gapB-pgk* transcription. *J. Bacteriol.* 180:6476–6483.

[7] Saier MH, Jr. (1996). Cyclic AMP-independent catabolite repression in bacteria. *FEMS Microbiol.Lett.* 138: 97-103.

[8] Ramseier TM. (1996). Cra and the control of carbon flux via metabolic pathways. *Res. Microbiol.* 147:489-497.

[9] Ramseier TM, Negre D, Cortay JC, Scarabel M, Cozzone AJ, Saier MH, Jr. (1993). In vitro binding of the pleiotropic transcriptional regulatory protein, FruR, to the fru, pps, ace, pts and icd operons of *Escherichia coli* and *Salmonella typhimurium*. *J. Mol. Biol.* 234:28-44.

[10] Ramseier TM, Bledig S, Michotey V, Feghali R, Saier MH, Jr. (1995). The global regulatory protein FruR modulates the direction of carbon flow in *Escherichia coli*. *Mol. Microbiol.* 16:1157-1169.

[11] Bledig SA, Ramseier TM, Saier MH, Jr. (1996). FruR mediates catabolite activation of pyruvate kinase (pykF) gene expression in *Escherichia coli*. *J. Bacteriol.* 178:280-283.

[12] Ow DSW, Lee RMY, Nissom PM, Philp R, Oh SKW, Yap MGS. (2007). Inactivating FruR global regulator in plasmid-bearing *Escherichia coli* alters metabolic gene expression and improves growth rate. *J. Biotechnol.* 131:261-269.

[13] Cortay JC, Negre D, Scarabel M, Ramseier TM, Vartak NB, Reizer J, Saier MH, Jr., Cozzone AJ. (1994). In vitro asymmetric binding of the pleiotropic regulatory protein, FruR, to the ace operator controlling glyoxylate shunt enzyme synthesis. *J. Biol. Chem.* 269:14885-14891.

[14] Negre D, Bonod-Bidaud C, Geourjon C, Deleage G, Cozzone AJ, Cortay JC. (1996). Definition of a consensus DNA-binding site for the *Escherichia coli* pleiotropic regulatory protein, FruR. *Mol. Microbiol.* 22:257-266.

[15] Shimada T, Fujita N, Maeda M, Ishihama A. (2005). Systematic search for the Cra-binding promoters using genomic SELEX system. *Genes to cells.* 10:907-918.

[16] Bagg A. Neilands JB. (1987). Ferric uptake regulation protein acts as a repressor, employing Iron(II) as a cofactor to bind the operator of an Iron transport operon in *Escherichia coli*. *Biochemisty.* 26:5471-5477.

[17] Vassinova N. Kozyrev D. (2000). A method for direct cloning of Fur-regulated genes: identification of seven new Fur-regulated loci in *Escherichia coli*. *Microbiol.* 146:3171-3182.

[18] Unden G. Schirawski J. (1997). The oxygen-responsive transcriptional regulator FNR of *Escherichia coli*: the search for signals and reactions. *Mol. Microbiol.* 25:205-210.

[19] Green J, Scott C. Guest JR. (2001). Functional versatility in the CRP-FNR superfamily of transcription factors: *FNR and FLP.* 44:1-34.

[20] Guest JR, Green J, Irvine AS, Spiro S. (1996). The FNR modulon and FNR regulated gene expression. In: Lin ECC, Lynch AS, (editors). Regulation of gene expression in *Escherichia coli*. New York: Chapman & Hall. 317–342.

[21] Shalel-Levanon S, San KY, Bennett GN. (2005). Effect of ArcA and FNR on the expression of genes related to the oxygen regulation and the glycolysis pathway in *Escherichia coli* under microaerobic growth conditions. *Biotechnol. Bioeng.* 92:147-159.

[22] Oh, MK, Liao JC. (2000).Gene expression profiling by DNA microarray and metabolic fluxes in *Escherichia coli*. *Biotechnol. Prog.* 16:278-286.

[23] Kabir MM, Shimizu K. (2003). Gene expression patterns for metabolic pathway in pgi knockout *Escherichia coli* with and without phb genes based on RT-PCR. *J. Biotechnol.* 105:11-31.

[24] Ow DSW, Nissom PM, Philp R, Oh SKW, Yap MGS. (2006). Global transcriptional analysis of metabolic burden due to plasmid maintenance in *Escherichia coli* DH5α during batch fermentation. *Enzyme Microb.Technol.* 39:391-398.

[25] Romeo T. (1998). Global regulation by the small RNA-binding protein CsrA and the non-coding RNA molecule CsrB. *Mol. Microbiol.* 29:1321-1330.

[26] Liu M, Yang H, Romeo T. (1995). The product of the pleiotropic gene *Escherichia coli* csrA modulates glycogen biosynthesis via effects on mRNA stability. *J. Bacteriol.* 177:2663-2672.

[27] Liu M, Romeo T. (1997). The global regulator CsrA of *Escherichia coli* is a specific mRNA-binding protein. *J. Bacteriol.* 179:4639-4642.

[28] Baker CS, Morozoc I, Suzuki K, Romeo T, Babitzke P. (2002). CsrA regulates glycogen biosynthesis by preventing translation of glgC in *Escherichia coli*. *Mol. Microbiol.* 44:1599-1610.

[29] Wei B, Brun-Zinkernagel A. Simecka JW. Prub BM, Babitzke P, Romeo T. (2001). Positive regulation of motility and flhDC expression by the RNA-binding protein CsrA of *Escherichia coli*. *Mol. Microbiol.* 40:245-256.

[30] Sabnis NA, Yang H, Romeo T. (1995). Pleiotropic regulation of central carbohydrate metabolism in *Escherichia coli* via the gene csrA. *J. Biol. Chem.* 270:29096-29104.

[31] Fraekel DG. (1996). Glycolysis. In: Neidhardt FC, ed. Escherichia coli and Salmonella. Washington, D.C.: American Society for Microbiology. 189-198.

[32] Wachi M, Umitsuki G, Shimizu M, Takada A, Nagai K. (1999). *Escherichia coli* cafA gene encodes a novel RNase, designated as RNase G, involved in processing of the 5' end of 16S rRNA. 259:483-8.

[33] Lee K, Bernstein JA, Cohen SN. (2002). RNase G complementation of rne null mutation identifies functional interelationships with RNase E in *Escherichia coli*. *Mol. Microbiol.* 43:1445-1456.

[34] Kaga N, Umitsuki G, Nagai K and Wachi M. (2002). RNase G-dependent degradation of the eno mRNA encoding a glycolysis enzyme enolase in *Escherichia coli*. *Biosci. Biotechnol. Biochem.* 66:2216-2220.

[35] Sakai T, Nakamura N, Umitsuki G, Nagai K, Wachi M. (2007). Increased production of pyruvic acid by *Escherichia coli* RNase G mutants in combination with cra mutations. *Appl. Microbiol. Biotechnol.* 76:183-192.

[36] Stülke J, Hillen W. (2000). Regulation of carbon catabolism in *Bacillus* species. *Annu. Rev. Microbiol.* 54:849-880.

[37] Saier MH, Jr., Chauvaux S, Deutscher J, Reizer J, Ye JJ. (1995) Protein phosphorylation and regulation of carbon metabolism in gram-negative versus gram-positive bacteria. *Trends Biochem. Sci.* 20:267-271.

[38] Reizer J, Hoischen C, Titgemeyer F, Rivolta C, Rabus R, Stulke J, Karamata D, Saier MH, Jr., Hillen W. (1998). A novel protein kinase that controls carbon catabolite repression in bacteria. *Mol. Microbiol.* 27:1157-1169.

[39] Galinier A, Haiech J, Kilhoffer MC, Jaquinod M, Stulke J, Deutscher J, Martin-Verstraete I. (1997). The *Bacillus subtilis* crh gene encodes a HPr-like protein involved in carbon catabolite repression. *Proc. Natl. Acad. Sci. U.S.A.* 94:8439-8444.

[40] Deutscher J, Kuster E, Bergstedt U, Charrier V, Hillen W. (1995). Protein kinase-dependent HPr/CcpA interaction links glycolytic activity to carbon catabolite repression in gram-positive bacteria. *Mol. Microbiol.* 15:1049-1053.

[41] Tobisch S, Zuhlke D, Bernhardt J, Stulke J, Hecker M. (1999). Role of CcpA in regulation of the central pathways of carbon catabolism in *Bacillus subtilis*. *J. Bacteriol.* 181:6996-7004.

[42] Ludwig H, Homuth G, Schmalisch M, Dyka FM, Hecker M, Stulke J. (2001). Transcription of glycolytic genes and operons in *Bacillus subtilis*: evidence for the presence of multiple levels of control of the gapA operon. *Mol. Microbiol.* 41:409-422.

[43] Fillinger S, Boschi-Muller S, Azza S, Dervyn E, Branlant G, Aymerich S. (2000). Two glyceraldehyde-3-phosphate dehydrogenases with opposite physiological roles in a nonphotosynthetic bacterium. *J. Biol. Chem.* 275:14031-14037.

[44] Henkin TM, Grundy FJ, Nicholson WL, Chambliss GH. (1991). Catabolite repression of alpha-amylase gene expression in *Bacillus subtilis* involves a trans-acting gene product homologous to the *Escherichia coli* lacl and galR repressors. *Mol. Microbiol.* 5:575-584.

[45] Moreno MS, Schneider BL, Maile RR, Weyler W, Saier MH, Jr. (2001). Catabolite repression mediated by the CcpA protein in *Bacillus subtilis*: novel modes of regulation revealed by whole-genome analyses. *Mol. Microbiol.* 39:1366-1381.

[46] Miwa Y, Nakata A, Ogiwara A, Yamamoto M, Fujita Y. (2000). Evaluation and characterization of catabolite-responsive elements (cre) of *Bacillus subtilis*. *Nucleic Acids Res.* 28:1206-1210.

[47] Weickert MJ, Chambliss GH. (1990). Site-directed mutagenesis of a catabolite repression operator sequence in *Bacillus subtilis*. *Proc. Natl. Acad. Sci. U.S.A.* 87:6238-6242.

[48] Schumacher MA, Allen GS, Diel M, Seidel G, Hillen W, Brennan RG. (2004). Structural basis for allosteric control of the transcription regulator CcpA by the phosphoprotein HPr-Ser46-P. *Cell.* 118:731-741.

[49] Jones BE, Dossonnet V, Kuster E, Hillen W, Deutscher J, Klevit RE. (1997). Binding of the catabolite repressor protein CcpA to its DNA target is regulated by phosphorylation of its corepressor HPr. *J. Biol. Chem.* 272:26530-26535.

[50] Ludwig H, Rebhan N, Blencke HM, Merzbacher M, Stulke J. (2002). Control of the glycolytic gapA operon by the catabolite control protein A in *Bacillus subtilis*: a novel mechanism of CcpA-mediated regulation. *Mol. Microbiol.* 45:543-553.

[51] Faires N, Tobisch S, Bachem S, Martin-Verstraete I, Hecker M, Stulke J. (1999). The catabolite control protein CcpA controls ammonium assimilation in *Bacillus subtilis*. *J. Mol. Microbiol. Biotechnol.* 1:141-148.

[52] Doan T, Aymerich S. (2003). Regulation of the central glycolytic genes in *Bacillus subtilis*: binding of the repressor CggR to its single DNA target sequence is modulated by fructose-1,6-bisphosphate. *Mol. Microbiol.* 47:1709-1721.

[53] Zorrilla S, Chaix D, Ortega A, Alfonso C, Doan T, Margeat E, Rivas G, Aymerich S, Declerck N, Royer CA. (2007a). Fructose-1,6-bisphosphate acts both as an inducer and as a structural cofactor of the central glycolytic genes repressor (CggR). *Biochemistry.* 46:14996-15008.

[54] Zorrilla S, Doan T, Alfonso C, Margeat E, Ortega A, Rivas G, Aymerich S, Royer CA, Declerck N. (2007b). Inducer-modulated cooperative binding of the tetrameric CggR repressor to operator DNA. *Biophys. J. 92*:3215-3227.

[55] Rezacova P, Kozisek M, Moy SF, Sieglova I, Joachimiak A, Machius M, Otwinowski Z. (2008). Crystal structures of the effector-binding domain of repressor Central glycolytic gene Regulator from *Bacillus subtilis* reveal ligand-induced structural changes upon binding of several glycolytic intermediates. *Mol. Microbiol.* 69:895-910.

[56] Doan T, Martin L, Zorrilla S, Chaix D, Aymerich S, Labesse G, Declerck N. (2008). A phospho-sugar binding domain homologous to NagB enzymes regulates the activity of the central glycolytic genes repressor. *Proteins.* 71:2038-2050.

[57] Meinken C, Blencke HM, Ludwig H, Stulke J. (2003). Expression of the glycolytic gapA operon in *Bacillus subtilis*: differential syntheses of proteins encoded by the operon. *Microbiol.* 149:751-761.

[58] Servant P, Le Coq D, Aymerich S. (2005). CcpN (YqzB), a novel regulator for CcpA-independent catabolite repression of *Bacillus subtilis* gluconeogenic genes. *Mol. Microbiol.* 55:1435-1451.

[59] Tannler S, Fischer E, Le Coq D, Doan T, Jamet E, Sauer U, Aymerich S. (2008a). CcpN controls central carbon fluxes in *Bacillus subtilis*. *J. Bacteriol.* 190:6178-6187.

[60] Licht A, Preis S, Brantl S. (2005). Implication of CcpN in the regulation of a novel untranslated RNA (SR1) in *Bacillus subtilis*. *Mol. Microbiol.* 58:189-206.

[61] Licht A, Brantl S. (2006). Transcriptional repressor CcpN from *Bacillus subtilis* compensates asymmetric contact distribution by cooperative binding. *J. Mol. Biol.* 364:434-448.

[62] Licht A, Golbik R, Brantl S. (2008). Identification of ligands affecting the activity of the transcriptional repressor CcpN from *Bacillus subtilis*. *J. Mol. Bi*ol. 380:17-30.

[63] Tannler S, Zamboni N, Kiraly C, Aymerich S, Sauer U. (2008b). Screening of *Bacillus subtilis* transposon mutants with altered riboflavin production. *Metab. Eng.* 10:216-226.

[64] Daran-Lapujade P, Jansen MLA, Daran, JM, Van Gulik W, de Winde JH, Pronk JT. (2004). Role of transcriptional regulation in controlling fluxes in central carbon metabolism of *Saccharomyces cerevisiae*. *J. Biol. Chem.* 279:9125-9138.

[65] Fraenkel, DG. (2003). The top genes on the distance from transcript to function in yeast glycolysis. *Curr. Opin. Microbiol.* 6, 198-201.

[66] Daran-Lapujade P, Rossell S, Van Gulik W, Luttik MAH, de Groot MJL, Slijper M, Heck AJL, Daran JM, de Winde JH, Westerhoff HV, Pronk JT, Bakker BM. (2007). The fluxes through glycolytic enzymes in *Saccharomyces cerevisiae* are predominantly regulated at posttranscriptional levels. *Proc. Natl. Acad. Sci. U.S.A.* 104:15753-15758.

[67] Clifton D, Weinstock SB, Fraenkel DG. (1978). Glycolysis mutants of *Saccharomyces cerevisiae*. *Genetics*. 88:1-11.

[68] Baker HV. (1986). Glycolytic gene expression in *Saccharomyces cerevisiae*: nucleotide sequence of GCR1, null mutants, and evidence for expression. *Mol. Cell. Biol.* 6:3774-3784.

[69] Huie MA, Scott EW, DrazinicCM, Lopez MC, Hornstra IK, Yang TP, Baker HV (1992). Characterization of the DNA-binding activity of GCR1: in vivo evidence for two GCR1-binding sites in the upstream activating sequence of TPI of *Saccharomyces cerevisiae*. *Mol. Cell. Biol.* 12:2690-2700.

[70] Chambers A, Packham EA, Graham IR. (1995). Control of glycolytic gene expression in the budding yeast (*Saccharomyces cerevisiae*). *Current genetics*. 29:1-9.

[71] Baker HV. (1991). GCR1 of *Saccharomyces cerevisiae* encodes a DNA binding protein whose binding is abolished by mutations in the CTTCC sequence motif. *Proc. Natl. Acad. Sci. U.S.A.* 88:9443-9447.

[72] Graham IR, Haw RA, Spink KG, Halden KA, Chambers A. (1999). In vivo analysis of functional regions within yeast Rap1p. *Mol. Cell. Biol.* 19:7481-7490.

[73] Yu L, Sabet N, Chambers A, Morse RH. (2001). The N-terminal and C-terminal domains of RAP1 are dispensable for chromatin opening and GCN4-mediated HIS4 activation in budding yeast. *Journal of Biological Chemistry*. 276:33257-33264.

[74] Sasaki H, Kishimoto T, Mizuno T, Shinzato T, Uemura H. (2005). Expression of GCR1, the transcriptional activator of glycolytic enzyme genes in the yeast *Saccharomyces cerevisiae*, is positively autoregulated by Gcr1p. *Yeast*. 22:305-319.

[75] Uemura H, Fraenkel DG. (1990). gcr2, a new mutation affecting glycolytic gene expression in *Saccharomyces cerevisiae*. *Mol. Cell. Biol.* 10:6389-6396.

[76] Zeng X, Deminoff SJ, Santangelo GM. (1997). Specialized Rap1p/Gcr1p transcriptional activation through Gcr1p DNA contacts requires Gcr2p, as does hyperphosphorylation of Gcr1p. *Genetics*. 147:493-505.

[77] Deminoff SJ, Santangelo GM. (2001). Rap1p requires Gcr1p and Gcr2p homodimers to activate ribosomal protein and glycolytic genes, respectively. *Genetics*. 158:133-143.

[78] Nishi K, Park CS, Pepper AE, Eichinger G, Innis MA, Holland MJ. (1995). The GCR1 requirement for yeast glycolytic gene expression is suppressed by dominant mutations in the SGC1 gene, which encodes a novel basic- helix-loop-helix protein. *Mol. Cell. Biol.* 15:2646-2653.

[79] Sato T, Lopez MC, Sugioka S, Jigami Y, Baker HV, Uemura H. (1990). The E-box DNA binding protein Sgc1p suppresses the gcr2 mutation, which is involved in transcriptional activation of glycolytic genes in *Saccharomyces cerevisiae*. *FEBS Letters*. 463:307-311.

[80] Klein CJ, Olsson L, NielsenJ. (1998). Glucose control in *Saccharomyces cerevisiae*: the role of Mig1 in metabolic functions. *Microbiol.* 144:13-24.

[81] Nehlin JO, Ronne H. (1990). Yeast MIG1 repressor is related to the mammalian early growth response and Wilms' tumour finger proteins. 9:2891-2898.

[82] Lundin M, Nehlin JO, Ronne H. (1994). Importance of a flanking AT-rich region in target site recognition by the GC box-binding zinc finger protein MIG1. *Mol. Cell. Biol.* 14:1979-1985.

[83] Ozcan S, Johnston M. (1995). Three different regulatory mechanisms enable yeast hexose transporter (HXT) genes to be induced by different levels of glucose. *Mol. Cell. Biol.* 15:1564-1572.

[84] Francisca R, Pilar H, Pascual SJ, Antonio P, Fernando M. (1998). Hexokinase PII has a double cytosolic-nuclear localisation in *Saccharomyces cerevisiae*. *FEBS Letters.* 425: 475-478.

[85] Rodriguez A, De La Cera T, Herrero P, Moreno F. (2001). The hexokinase 2 protein regulates the expression of the GLK1, HXK1 and HXK2 genes of *Saccharomyces cerevisiae*. *Biochem. J.* 355, 625-631.

[86] Gancedo, Juana M. (2008). The early steps of glucose signalling in yeast. *FEMS Microbiol. Rev.* 32:673-704.

[87] Ahuatzi D, Riera A, Pelaez R, Herrero P, Moreno F. (2007). Hxk2 regulates the phosphorylation state of Mig1 and therefore its nucleocytoplasmic distribution. *J. Biol. Chem.* 282:4485-4493.

In: Glycolysis: Regulation, Processes and Diseases
Editor: Paul N. Lithaw
ISBN: 978-1-60741-103-1
© 2009 Nova Science Publishers, Inc.

Chapter VII

Blood Lactate Concentrations, Resistive Force Selection and High Intensity Cycle Ergometry: Metabolic Implications and Associations with Running Ability

Julien Steven Baker[1*] and Bruce Davies[2]

[1]Chair of Sport and Exercise Science, Division of Sport and Exercise Sciences, School of Science, University of the West of Scotland, Hamilton Campus, Almada Street, Hamilton. ML3 0JB

[2]Health and Exercise Science Research Laboratory, School of Applied Science, University of Glamorgan, Pontypridd, Wales, CF37 1DL.

Abstract

The purpose of this study was to analyse values generated during 30 s of high intensity cycle ergometry exercise when cradle resistive forces were calculated from total - body mass (TBM) or fat - free mass (FFM). A further aim was to compare the power values generated with performance indices recorded during maximal running performance on a modified multi stage fitness test and to validate the running test as a measure of anaerobic performance. Body density was calculated using underwater weighing procedures. Fat mass was estimated from body density values. Significant differences ($P < 0.01$) were observed between the TBM and FFM protocols for peak power output (PPO; 1264 ± 156W vs. 1366 ± 177W respectively). Significant differences ($P < 0.01$) were also recorded between the TBM and FFM protocols for resistive force selection and pedal revolutions (7.3 ± 1.2 vs. 6.2 ± 1.1 kg; 136 ± 8.7 vs. 144 ± 7.6 rpm respectively). There were no differences ($P > 0.05$) recorded between

[*] To whom correspondence should be addressed. Professor Julien Steven Baker; Email jsbaker@glam.ac.uk; Tel 01443 482972; Fax 01443 482285

mean power output (MPO) or fatigue index (FI %). Values recorded for the running test were 71.4 ± 7.5 s. Significant ($P < 0.01$) linear relationships were found between PPO and running times for both the TBM and FFM protocols with more of the variance accounted for during the FFM protocol. Blood lactate concentrations increased significantly from rest to 5 min post exercise for all three experimental conditions and were highly correlated ($P < 0.01$). Results from the study suggest that higher PPO values are obtainable when resistive forces used in high intensity cycle ergometry exercise reflect lean tissue mass. Also, the running test proved to be a viable measure for the quantification of high intensity running performance during periods of intense work.

Keywords: Running, cycle ergometry, body mass.

Introduction

Coaches, trainers and athletes are continually searching for optimum ways of identifying key elements that complement, and effect athletic outcomes. High intensity performances which principally involve short bursts of heavy exercise, such as sprinting or jumping rely predominantly on the immediate (ATP-PC) and short term (anaerobic glycolysis) energy production systems. The ability to utilise the high energy phosphate stores very quickly may be considered as one aspect of "high intensity power". While the total amount of energy available to perform work in a given energy system is referred to as the capacity of the system [4]. Individual differences in power production may be the result of greater muscle mass, or a greater proportion of fast twitch fibres that posses higher ATP-PC enzymatic activity [8]. Procedures for measuring and quantifying high intensity performances are varied. Such procedures have ranged from simple field tests, such as sprinting and jumping, to laboratory techniques comprising various modes of exercise, e.g. treadmill sprinting [5], stair climbing [13], vertical jumping, cycle ergometry [14], and various isokinetic measurements. Development of a 30 s cycle ergometer test [1] has enabled the measurement of the peak, mean and end power outputs, while observing fatigue profiles during exercise of maximum intensity. During the computation of resistive forces used in the assessment of power during high intensity cycle ergometer exercise, the assumption has been that the relationship between total - body mass (TBM) and fat - free mass (FFM) is the same. Recent research in our laboratory [2] has demonstrated that greater peak power outputs (PPO) are obtainable when resistive forces reflect the lean tissue component of body composition. Baker et al, [2] further demonstrated that variations in body composition between subjects may under or over estimate the resistive forces used in high intensity cycle ergometry when the forces are based on TBM computations. This may lead not only to spurious calculations of power output during high intensity cycle ergometry, but could also effect the validity and reliability of high intensity exercise field tests that have been validated in conjunction with high intensity cycle ergometry as the criterion measure.

The aim of this study was to examine any observed differences in high intensity friction loaded cycle ergometry power profiles using a TBM and FFM resistive force selection procedure and to investigate possible relationships with high intensity running performance using a modified aerobic shuttle run course as a measure of anaerobic ability. A further aim was to validate the running test as a measure of anaerobic performance.

Subjects and Experimental Design

Male university soccer players (n = 16) volunteered as subjects. Mean (± SD) for age, body mass, stature and % fat of the group can be found in table 1.

Prior to testing, all subjects were habituated to the experimental procedures, tested at the same time of day and were informed that they were free to withdraw from the experiment at any time. Prior to the investigation ethical procedures were approved by the university ethics committee and all subjects read and signed an informed consent form. A minimum of two rest days (no physical activity) preceded each test, and subjects attended the laboratory following an overnight fast in an attempt to control the influence of diet on performance.

Table 1. Mean ± SD for physiological characteristics of subjects

Variable	Mean ± SD
Age (yrs)	23.3 ± 2.1
Stature (cms)	183.5 ± 7.7
Mass (Kg FFM)	74.6 ± 9.1
Mass (Kg TBM)	87.2 ± 11.2
Fat %	12.5 ± 3.4

Anthropometric Measures

Body mass, stature and body composition was determined using a calibrated balanced weighing scale (Seca, UK), stadiometer (Seca, UK) and underwater weighing respectively. Nude body mass was measured to the nearest 0.1 kg and stature to 0.1 cm. Body density was assessed using underwater measures as described previously [3]. Relative body fat was estimated from body density [18]. Residual lung volume was measured using the simplified oxygen re-breathing method [22]. FFM was determined by subtracting fat mass from TBM.

Terminology

Throughout the study peak power output (PPO) refers to the greatest value for power recorded during the test. Mean power output (MPO) refers to the average power output during the 30 s test period. Fatigue index refers to the decrease in power over test duration, and is expressed as a percentage (FI %). PR refers to the highest pedal revolution recorded during the test.

Force Velocity Test

Prior to the 30 s cycle ergometer test a force velocity experiment was performed with one weeks rest between protocols to determine optimal resistive forces for TBM and FFM. The test consisted of six short maximal sprints (6-8s) against the following randomly assigned resistive forces: 70, 75, 80, 85, 90 and 95 $g.kg^{-1}$. Successive exercise bouts were separated by a 5 min rest period. The load that produced the highest PPO value for both the TBM and FFM protocol was considered to be optimal and was used in the 30 s cycle ergometer test. The optimal load selected for both the TBM and FFM protocol was verified using a test-retest method. Care was taken to ensure that problems outlined by Heiser, [9] concerning resistive force transmission to the flywheel during the exercise period were minimal. Therefore, resistive forces exceeding 9 kg were not used. On completion of the force velocity tests, subjects were again assigned in a random fashion, a week later, to the running test or to the remaining cycle ergometer TBM or FFM protocol. Rest periods of one week were observed between the three experimental conditions.

Cycle Ergometer Test Protocol

A cycle ergometer (Monark 864) was calibrated prior to warm up and data collection. The calibration procedure followed the guidelines for friction loaded cycle ergometers outlined by Coleman, [6]. The same calibration procedures were followed for the force velocity tests and 30 s cycle ergometer protocols.

Saddle heights were adjusted individually to accommodate partial flexion of the knee between 170° to 175° (with 180° denoting a straight leg position) in middle dead centre during the down stroke. Feet were firmly supported by toe clips and straps, and subjects were instructed to remain seated during the test. All subjects were verbally encouraged to perform maximally during testing, and all performed a standardised 5 minute warm up prior to data collection following procedures outlined by Jaskolska et al, [10]. Subjects were given a rolling start at approximately 60 rpm followed by a three second count down after which individual loads were applied and data capture initiated simultaneously.

Values for PPO, MPO, FI %, and PR were determined from flywheel revolutions using an inertia corrected computer programme [6]. Data transfer was made possible using a suitably mounted sensor unit and power supply attached to the fork of the ergometer.

The sampling frequency of the sensor was 18.2 Hz. Validity and reliability of the cycle ergometer as a test of anaerobic power has been reported as $r = 0.93$ [15].

Running Test

Subjects were required to run maximally on a modified aerobic shuttle run course. The original distance of the course was 20 m as outlined by Leger et al, [11]. The running test protocol comprised of running between two markers placed 15 m apart in a sports hall, at speed increases of 0.28 $m.s^{-1}$ each minute.

A reduction in distance of 5 m was implemented as a result of pilot study findings. The high intensity shuttle run start point was located on a line identified clearly by markers.

Before commencement of the test, subjects were given a familiarisation trial of five low intensity runs simulating experimental conditions. The protocol consisted of running to the first marker, turning and running again in the opposite direction in time with beeps emitted from a modified audiocassette tape recorded at twice the speed of the original [11]. Failure to make the line in the prescribed time on two occasions resulted in disqualification from the test. Turning procedures for all subjects were standardised prior to data collection. Before testing, the audio signal used in the test was checked for accuracy and reliability. The time taken from the start of the test to the disqualification point was recorded digitally by the same experimenter. Reliability for the running test was established using a test re-test method prior to data collection ($P < 0.01$). Subjects were required to perform maximally on each occasion.

The total time taken from the commencement of the test to volitional exhaustion (when the subjects failed to reach the line in the prescribed time) was taken as the criterion measure. Heart rate recordings for each subject were measured pre and post exercise using a short range telemetry system (Sport Tester 3000, Polar Electro, Finland).

Capillary Blood Sampling

Duplicate blood samples were collected at the same time of day and by the same investigator in an attempt to control for biological and between subject variation [16]. To help control for plasma volume changes, all resting samples were taken following 30 minutes of supine rest. The 5 min post exercise samples were taken with subjects placed in a supine position on a clinical couch to minimise any risk of fainting. This procedure was followed for all three protocols. All capillary blood samples were corrected for plasma volume changes using the equations of Dill and Costill [7].

Samples from the right ear lobe were collected using a capillary tube. All blood samples were analysed immediately post retrieval. Blood lactate concentrations were determined using an analox (P-LM5) lactate analyser. Haematocrit was analysed by the Hawksley micro haematocrit reader. The haematocrit reader was cleaned with mediswabs between each subject. haemoglobin was collected in a haemocue and measured with a photometer.

Statistical Analysis

Parametric statistical analysis was used in this study following conformation of a normal data distribution (SPSS). Paired t-tests were used to analyse differences between power indices, resistive force selection and pedal revolutions recorded for the two cycle ergometer tests. Pearson's correlation analysis was utilised to identify the degree of linear relationship between the cycle ergometry protocols, blood lactate concentrations and running performance. Correlational analysis was also used to investigate any relationships between heart rate responses measured for the three tests. Significance was accepted at $P < 0.05$.

Results

Physiological characteristics of the group are given in table 1. Performance data for cycle ergometry and the running test are given in table 2. Significant differences were found between PPO for both the TBM and FFM protocols ($P < 0.01$). There were no differences observed between MPO and FI between the cycle ergometry experimental conditions. Differences were observed between resistive forces and pedal revolutions ($P < 0.01$) when the TBM and FFM protocols were compared. High correlations were recorded between TBM and FFM for PPO and shuttle run times ($P < 0.01$). Interestingly, the FFM resistive force selection procedure accounted for more of the variance in performance when the running test was compared to both the TBM and FFM protocols ($r = 0.70$, $P < 0.01$ $R^2 = 50\%$ TBM; $r = 0.88$, $P < 0.01$ $R^2 = 77\%$ FFM respectively). Blood lactate concentrations for all tests were highly correlated ($P < 0.01$) and increased significantly ($P < 0.01$) from rest to post exercise values for all three experimental conditions. Exercise heart rate values of 176 ± 8 b.min^{-1} (TBM), 179 ± 6 b.min^{-1} (FFM) and 184 ± 6 b.min^{-1} were recorded for cycle ergometry, and the running test respectively and were also highly correlated ($P < 0.01$).

Table 2. Mean ± SD values for power output measurements, cradle forces and pedal revolutions recorded for both cycle ergometer protocols. Running results, heart rates and post exercise blood lactate concentrations are also given

Variable	TBM	FFM	Running Test
PPO (W)	1264 ± 156	1366 ± 177*	#
MPO(W)	878 ± 107	857 ± 119	#
Pedal/revs (rpm)	136 ± 8.7	144 ± 7.6*	#
FI%	35 ± 5	38 ± 8	#
B/L(mmol.l^{-1})p/ex B/L(mmol.l^{-1})rest	12.2 ± 1 0.8 ± 0.5	12.9 ± 1.7 1.1 ± 0.8	13.6 ± 1.8 0.8 ± 0.3
M/cradle (Kg)	7.3 ± 1.2	6.2 ± 1.1*	#
HR (bpm)post HR (bpm)rest	176 ± 8 72 ± 6	179 ± 6 70 ± 8	184 ± 6 70 ± 4
Running Time (s)	#	#	71.4 ± 7.5

* P < 0.01 Indicates differences between the TBM and FFM protocol.

Discussion

Significant differences were found between PPO for both the TBM and FFM protocol ($P < 0.01$). The lighter resistive forces used during the FFM protocol resulted in a significantly greater pedal velocity when compared to TBM ($P < 0.01$). The increase in pedal velocity contributed to the greater power outputs observed for the FFM protocol. This was consistent with resistive forces being significantly lighter during the FFM experimental condition ($P < 0.01$). Dotan et al, [8] reported that high power outputs during cycle ergometry were due to the optimisation of the resistive force. This is true in the present study. However, the higher

PPO values obtained for FFM resulted from an increase in pedal revolutions and a decrease in resistive force. These results are interesting when we consider that there were no differences observed for FI % between the TBM and FFM experimental conditions. The lack of significance observed may be the result of manipulations in resistive force selection that result in pedal velocity differences between the two protocols (higher resistive forces for TBM, and higher pedal velocities for FFM). These findings are in agreement with the suggestions of Wilkie, [21] and Van Mil et al, [20] who stated that force should be matched to the capacity of active muscle in order to exploit the full force velocity relationship. Significant linear relationships were recorded between the peak power output values for both the TBM and FFM protocols and the times recorded for the high intensity shuttle run test ($P < 0.01$).

These findings are in agreement with Cheetham et al, [5] who found strong linear relationships between running ability and high intensity PPO. More of the variation in performance was accounted for during the FFM protocol when running performances and both the TBM and FFM protocols were compared ($r = 0.70$, $P < 0.01$ $R^2 = 50\%$ TBM; $r = 0.88$, $P < 0.01$ $R^2 = 77\%$ FFM). The findings suggest that the FFM cycle ergometer protocol is more representative of high intensity performance than values that are inclusive of the fat component of body composition.

The design of the high intensity shuttle run course which includes slowing down at the end of each stage to facilitate turning, may weaken the relationship observed with the power outputs obtained during the cycle ergometry tests. However, strong correlations were recorded for both the TBM and FFM method of resistive force selection which support the use of the test as a measure of high intensity running ability. During the quantification of high intensity performance the metabolic contribution of the aerobic system to energy supply during this type of activity has to be considered. Studies by Smith et al, [19] have suggested an aerobic contribution of 16% during a 30 s high intensity cycle ergometer test. Although inevitable, this high contribution is not desirable and may compromise the assessment of anaerobic ability. The optimisation procedure for FFM may have maximised the anaerobic component during cycle ergometer exercise and minimised the aerobic influence by virtue of the faster pedal velocities obtained. The higher power outputs may be the result of an increased utilisation of ATP- PC or increased contribution from anaerobic glycolysis or both. It is unlikely that the increases in power observed for the FFM protocol was attributable to aerobic metabolism as the increase occurred in the first few seconds of exercise. Serrese et al, [17] and Smith et al, [19] have both suggested that aerobic metabolism plays a minor role in energy supply during high intensity exercise of 30 s duration until the 12th second.

However, the increase in contribution of aerobic metabolism to anaerobic performances as the exercise proceeds, may have contributed to energy supply during both the running test and the cycle ergometer tests. This may be true in spite of the fact that the greater intensities of performance were observed in the early stages of the cycle ergometer tests and the later stages of the running test. In addition, aerobic metabolism may have been dominant in the early phases of the running test and may represent the low intensity performance required at this early stage. The greater contribution from anaerobic metabolism would have progressively increased as the intensity became more related to anaerobic ability. This would arise when the speed required to run each level of the course increased from the previous level. The values recorded for the high intensity shuttle run test (71.4 ± 7.5 s) suggest that all subjects reached a similar level of performance. This indicates that the audio tape speed may have been emitting beeps at a pace that was not attainable for most subjects at the volitional fatigue point. It also suggests that the exercise group may have possessed similar physiological characteristics and were of a comparable training status. During running performances power outputs can be divided into two components, a vertical component for lifting the centre of gravity and a horizontal component for propulsion. The vertical component was not measured in this study. However, horizontal propulsive power is about 70% of the total external power at maximal running velocity [10]. To some degree the running test values may be underestimated because velocity, force and power all increase and decrease during each stride depending on the phase of running [10]. In spite of this, the strengths of the correlations obtained suggest that the FFM protocol represents a measure that is more related to high intensity running ability than the TBM protocol. Body mass is an important component to consider in the assessment of high intensity performance, as the findings of this study demonstrate. However, other factors such as training specificity and the fibre type distribution within the muscle may also contribute to force generation over short time periods and need consideration [12]. There were no differences in blood lactate concentrations pre or 5 min post exercise for the three experimental conditions. The TBM and FFM cycle ergometer test durations were the same and the lack of statistical difference between the MPO values indicate that the magnitude of glycolytic activation was similar for both protocols. During the running test, despite the duration of performance being more than twice the magnitude of the cycle ergometer protocols, in the early stages of the test, the speeds observed are of a low intensity, and therefore may result in minimal lactate accumulation. Lactate production may exceed the rate of removal at higher levels of performance and accumulation may occur in the later stages as both speed and intensity of running increase. In addition, it is worth noting that both the TBM and FFM cycle ergometer protocols utilised maximal resistive forces which may have contributed to the lactate values observed. Training status, muscle fibre composition and strength may also affect lactate production and removal. Heart rate values observed for the three protocols were highly correlated suggesting that the work intensities of each are related causing similar cardiovascular responses. The correlations obtained between the high intensity shuttle run test and both the TBM and FFM cycle ergometry protocols indicate that the shuttle may be valuable in quantifying high intensity performance ability, when sophisticated laboratory measures are not available. The high intensity shuttle run test may be useful to individual athletes, and performers involved in most team sports, where the nature of the activity requires the assessment of high intensity running performance. The shuttle may also be used as a training aid to evaluate the success of training programmes where the emphasis is on the development of high intensity ability. In addition to the shuttle test being easy to administer, linear relationships obtained from test re-test procedures indicate that the test is both reproducible and reliable. The findings of this study indicate that a modified aerobic shuttle run test as a measure of anaerobic ability provided a viable protocol for the quantification of high intensity exercise during periods of short intensive work. The relative strengths of the correlations observed between the running tests and the different cycle ergometer protocols indicate that the FFM protocol was more related to running ability than the TBM procedure.

Also, the cycle ergometer optimisation protocols examined, produced greater power outputs for FFM when compared to TBM. Loading procedures that relate to the active muscle tissue utilised during this type of exercise may need to be explored in preference to protocols that include both lean and fat masses.

References

[1] Aylon A, Inbar O, and Bar – Or, O. *Relationships between explosive strength and anaerobic power*. In "International series on sport sciences". Vol 1. Biomechanics IV, University Press, Baltimore (Nelson, R.C. and Morehouse, C.A.), 1974; 572 - 577.

[2] Baker JS, Bailey D, and Davies B. The relationship between total body mass, fat free mass and cycle ergometer power components of 20 s duration. *J. of Sci. and Med. in Sport*. 2001; 4: 1-9.

[3] Behnke AR, Wilmore JH. *Evaluation of body build and composition*. (Englewood Cliffs, N. J) Prentice Hall Inc., Hertfordshire. 1974 ; 20-24.

[4] Bouchard C, Taylor A, and Dulac J. *Testing anaerobic power and capacity*. In "Physiological Testing of The Elite Athlete". (J. D. Mcdougall, M. A. Wenger and J. H. Green eds) Human Kinetics, Champaign, Illinois. 1991; 176-214.

[5] Cheetham M, Boobis L, Brooks S, and Williams C. Human muscle metabolism during sprint running. *J. Appl. Physiol*. 1986; 61: 54-60.

[6] Coleman, S. "Corrected Wingate Anaerobic Test". Cranlea and Co, Sandpits Lane, Acacia Rd, Bournville, B'ham, U.K. 1996; 4-5.

[7] Dill DB, and Costill DL. Calculation of the percentage changes in volumes blood, plasma and red cells in dehydration. *J. of Appl. Physiol*. 1974; 37: 247-248.

[8] Dotan R, and Bar-Or O. Load optimisation for the Wingate anaerobic test. *Eur. J. Appl. Physiol*. 1983; 51: 409-417.

[9] Heiser K. Load optimization for peak and mean power output on the Wingate anaerobic test. Unpublished Masters Thesis. Arizona State University.1989.

[10] Jaskolska A, Goossens P, Veenstra B, Jaskolski A, and Skinner JS. Comparison of treadmill and cycle ergometer measurements of force – velocity relationships and power outputs. *Int. J. of Sports Med*. 1999; 20: 192 - 197.

[11] Leger L, Mercier D, Gadoury C, and Lambert J. The Multistage 20m shuttle run test for aerobic fitness. *J. of Sport Sci*. 1988; 6: 93-101.

[12] Manning GJ, Manning C, and Perrin D. Factor analysis of various anaerobic power tests. *J. of Sports Med. and Phys. Fitness*. 1988; 28: 138-144.

[13] Margaria R, Aghemo P, and Rovelli E. Measurement of muscular power in man (Anaerobic) *J. Appl. Physiol*. 1966; 21: 1662-1664.

[14] Mccartney N, Heigenhauser G, and Norman J. Power output and fatigue of human muscle in maximal cycling exercise. *J. Appl. Physiol*. 1983; 55: 218-224.

[15] Patton J, Murphy M, and Frederick F. Maximal power outputs during the Wingate anaerobic test. *Int. J. of Sports Med*. 1985; 6: 82 - 5.

[16] Reilly T, and Brooks GA. Investigation of circadian rhythms in the metabolic response to exercise. *Ergonomics*. 1982; 25: 1093-1197.

[17] Serresse O, Lortie C, Bouchard C, and Boulay MR. Estimation of the contribution of the various energy systems during maximal work of short duration. *Int. J. of Sports Med.* 1988; 9: 456 - 460.

[18] Siri WE. *Gross composition of the body*. In "Advances in biological and medical physics. IV" (Lawrence, J. H. and Tobias C. A.) New York Academic Press 1956; 239-280.

[19] Smith JC, and Hill DW. Contribution of energy systems during a Wingate power test. *Brit. J. of Sports Med.* 1991; 25:196-199.

[20] Van mil E, Schoeber N, Calvert R, and Bar – Or O. (1996): Optimisation of force in the Wingate test for children with a neuromuscular disease. *Med. and Sci. in Sport and Ex.* 1996; 28: 1087 - 1092.

[21] Wilkie DR. Man as a source of mechanical power. *Ergonomics*. 1960; 3: 1 - 8.

[22] Wilmore JH, Vodak PA, Parr RB, Girandola RN, and Billing JE. Further simplification of a method for determining residual lung volume. *Med. and Sci. in Sport and Exer.* 1980; 12: 216-218.

In: Glycolysis: Regulation, Processes and Diseases
Editor: Paul N. Lithaw
ISBN: 978-1-60741-103-1
© 2009 Nova Science Publishers, Inc.

Chapter VIII

Blood Lactate Concentrations Following Repeat Brief Maximal Intermittent Exercise in Man. Glycolytic Energy Supply and Influence of Plasma Volume Changes

Julien S. Baker[1,*], Christopher J. Retallick[2,†], Peter Reynolds[1], Bruce Davies[2,‡] and Robert A. Robergs[3,#]

[1]Chair of Sport and Exercise Science, Division of Sport and Exercise Sciences, School of Science, University of the West of Scotland, Hamilton Campus, Hamilton. ML3 0JB
[2] Health and Exercise Science Research Unit, Department of Science and Sport, University of Glamorgan, Pontypridd, Wales, CF37 1DL
[3] Exercise Physiology Laboratories, Exercise Science Program, Department of Physical Performance and Development, University of New Mexico, Albuquerque, New Mexico 87131, USA

Abstract

Background: Energy system interaction during repeated bouts of maximal activity is complex, and relatively little is known about different energy system contribution during exercise.

[*] Address for correspondence: Professor Julien S Baker; Division of Sport and Exercise Sciences, School of Science, University of the West of Scotland, Hamilton Campus, Almada Street, Hamilton. ML3 0JB CF37 1DL. Telephone: +44 (0) 1443 482972; Fax: +44 (0) 1443 2285; e-mail: jsbaker@glam.ac.uk
[†] cjretall@glam.ac.uk
[‡] bdavies1@glam.ac.uk
[#] rrobergs@unm.edu

Objective: The aim of the present study was to examine the contribution of anaerobic glycolysis to a repeat sprint protocol via the assessment of blood lactate concentration.

Research design and methods: Eight male, healthy subjects volunteered to participate in the study. The subjects performed eight 6-s sprints on a friction loaded cycle ergometer with a 60-s recovery period between each sprint. Plasma volume corrected blood samples were collected at rest (following 30 min in a supine position), following each sprint (within the first 10-s) and at 5 min post-exercise.

Results: The highest mean (MPO) and peak power output (PPO) was observed in the first and third sprint for both conditions (777.3 ± 142.2 W and 874.9 ± 175.6 W, respectively; see figure 1). Power outputs were maintained during the exercise period with no significant decreases observed between sprint 1 and eight ($P > 0.05$). In contrast, blood lactate concentrations increased throughout the successive sprint periods from a resting value of 0.67 ± 0.47 mmol/L, to a peak value of 7.5 ± 1.8 mmol/L, immediately following sprint 8 ($P < 0.05$) Plasma volume changes showed a gradual haemoconcentration after sprint two (-0.86 ± 5.94%), and approached a significant change from the resting value immediately after sprint eight (~9.5% haemoconcentration; $P < 0.05$).

Conclusions: The main findings of this study were that 60-s recovery from brief maximal exercise is sufficient to replenish the anaerobic energy stores (ATP-PC).and that anaerobic glycolysis plays a significant role in energy provision as exercise progressed.

Keywords: Intermittent exercise, Blood lactate, Power output.

Introduction

Sporadic bouts of all-out maximal exercise characterise most, if not all, popular recreational sports. Energy system interaction in this type of exercise is complex and relatively little is known for certain, about the proportionate contributions made by the three main energy producing processes (Gastin 2001) to adenosine triphosphate (ATP) re-synthesis during repetitive brief maximal exercise (RBME) (Gaitanos et al. 1993).

Blood lactate concentration continues to be utilised to offer insight into the anaerobic contribution to exercise (Jacobs 1986). Previous studies using blood lactate measurements during intermittent maximal exercise have suggested that short intervals of recovery (30-s) result in reduction of the muscle phosphocreatine (PCr) content, and subsequently place greater demand on glycogenolysis to provide ATP anaerobically during successive sprints (Holmyard et al.1988; Wooton and Williams 1983). It may seem surprising then that other studies using relatively similar work-to-rest ratios ranging from 10-15-s:15-40-s (work: rest ratio) have suggested that the contribution of glycogenolysis to the total energy demand was of minor importance (Saltin and Essen 1971). This disparity is probably due to differences in protocol, since the studies by Saltin and Essen did not utilise truly maximal exercise intensities. However, the study remains relevant since it identifies the complexity of data interpretation for exercise involving repeated work/rest intervals.

Gaitanos and co-workers (1993), utilising a protocol that involved ten maximal 6-s cycle sprints separated by 30-s of recovery between each sprint, showed a considerable reduction in the contribution of anaerobic glycolysis to ATP production. They suggested that during the last sprint, power output was mainly derived from PCr degradation and an increased aerobic metabolism. However, earlier findings from the same laboratory concluded that an identical protocol lead to an increasing demand on anaerobic glycolysis to maintain the rate of energy production (Wooton and Williams 1983). Clearly overall metabolic consequences of RBME are affected not only by the duration and workload of the exercise bout, but also of the duration of the recovery period (Holmyard et al. 1988). RBME protocols continue to be used in the assessment of muscular performance (McCartney et al. 1986) and as a training protocol to improve aerobic as well as anaerobic metabolism (Rodas et al. 2000). To date, the greatest interest has been placed on the concepts of metabolism (Hermansen and Vaage 1977; McCartney et al. 1986) and fatigue (Welsh et al. 2002). In addition, when dealing with single brief bouts (<10-s) of exercise in isolation the literature can confidently predict which system will predominate, the apparent disparity arises when the periods of exercise and recovery are no longer dealt with in isolation.

In an attempt to advance the research on high-intensity intermittent exercise, the present study employed a friction loaded cycle ergometer protocol in an attempt to examine the glycolytic contribution to 8 successive 6-s bouts of maximal exercise, with 60-s of rest between each sprint. The objectives of this study were to examine any changes in maximal dynamic power and fatigue at the end of each 6-s bout for the eight sprints. A further aim was to identify the associated metabolic changes at the end of each 6-s stage by monitoring blood lactate concentrations corrected for plasma volume shifts. The null hypothesis proposed for this experiment was that blood lactate concentration and power outputs would not be significantly different between any of the sampling periods.

Methods

Eight healthy male non-smokers volunteered to participate in the present study. All were moderately active individuals who regularly took part in team sport at an amateur level. Their age, height, weight and body fat were 25 ± 4 yrs, 179.3 ± 5.7 cm, 85.0 ± 9.5 kg and 16.5 ± 5.6 % (mean ± SD) respectively. Experimental design and procedures were granted ethical approval by the University. Informed consent was provided in accordance with the Helsinki convention, and was obtained from all participants after test procedures were thoroughly explained. Prior to any maximal physical exertion, all subjects were fully familiarised and reminded of their freedom to withdraw from the experiment at any time.

The participants were instructed to consume their normal diet on the day preceding the test. Additionally, subjects were asked to refrain from exercise for two days before the test and asked not to consume caffeine, alcohol, medications (non-essential) or supplements for one-day prior testing. In an attempt to maintain hydration levels, which are known to affect plasma volume (Kargotich et al. 1998), subjects were asked to consume adequate amounts of water, one day prior to the actual test and on the morning of the test (at least two hours before testing)

Laboratory tests

The familiarisation procedure was conducted one week prior to testing. On arrival, subjects were given a schematic representation of the testing procedure. This was explained in detail to all subjects, and any subsequent questions were answered in full.

Skinfold Measurement: chest, midaxillary, triceps, subscapular, abdomen, suprailiac and thigh skinfold measurements were taken in accordance with the Anthropometric Standardisation Reference Manual (Lohman et al. 1988). Measurements were carried out on the right hand side of the body, with Harpenden Skinfold callipers (British Indicators, RH15 9LB, England), and measured to the nearest 0.2mm. Generalised skinfold equations (Jackson, 1984), were then utilised to calculate body density, which, were then converted to percentage of body fat (%BF), using population specific formulas (Heyward, 1991). In addition, each participant completed one exact replica of the test protocol at a sub-maximal intensity, after completing a comprehensive medical questionnaire and giving full written informed consent.

Exercise tests: All laboratory testing took place between 9am and 12pm to minimise diurnal variation. Participants arrived for testing having fasted overnight (for at least a 10hr period) and were euhydrated. Exercise was performed on friction-loaded cycle ergometer (Monark 864E, Monark, Varberg, Sweden), described in detail previously (Lakomy 1986). Saddle heights and handle bar positions were individually adjusted to allow a slight bend at the elbow and knee joint when the arms and left-leg were extended; with the participant sat in an upright seated posture. Toe-clips were fitted and participants were reminded of the maximal nature of the test and of the importance of remaining seated throughout the test duration.

Following 15min of seated rest, subjects were warmed up for 3min at 60rpm against a constant resistance of 1.5kg. The test procedure consisted of eight 6-s maximal cycle sprints against a standard load of $75g.Kg^{-1}$; each sprint was separated by 60-s of seated static rest, and followed the same basic procedure. Subjects were informed verbally to begin slow, unloaded pedalling, 5-s before each sprint. All out pedalling (still unloaded) began three seconds prior sprinting in order to overcome the initial inertia and frictional resistance of the ergometer, then at 0-s the resistance was applied and the test began. Test procedures have been outlined previously (Smith and Hill, 1991). Heart rate, blood lactate, haemoglobin and haematocrit were sampled at rest (post 15min seated rest), immediately after each sprint and at 5min post exercise. In addition, the cycle ergometer was interfaced with a personal computer to allow analysis of peak power and mean power via inertia corrected computer software (Coleman, 1996).

Blood Analysis

Capillary blood samples from all subjects were drawn from the left ear lobe. Prior to the initial sample being collected each participant's ear lobe was first cleaned with an alcohol-based medical swab, the site was then allowed to dry for approximately one minute after which the first incision was made with a sterile lancet. The initial blood droplet was discarded.

Blood Lactate: intact whole blood was collected into 50μl fluoride/heparin/nitrite glass capillary tubes and mixed for around 30s. A 7μl sample was then drawn from the capillary tube with a plastic pipette and analysed with an Analox® GL5 L-Lactate analyser (Analox Instruments LTD, Hammersmith, London).

Haemoglobin and haematocrit: A 10μl sample of whole blood was drawn from the earlobe into a HemoCue® (HemoCue®, Ängelholm, Sweden) B-Haemoglobin microcuvette and then analysed photometrically using the modified azidemethemoglobin reaction. Blood for haematocrit analysis was drawn into 75μl heparinised capillary tubes (Hawksley, Sussex, England) adequately mixed and haematocrit was visually determined with an Hct meter following centrifuging at 3000 x G for 3 minutes. Capillary samples were corrected for plasma volume changes using the equations of Dill and Costill (1971).

Calibration Procedures

Blood lactate: A 7μl sample of aqueous 8.0mmol/L (72.1mg/dL) lactate standard was utilised to calibrate the Analox Analyser (Analox Instruments LTD, Hammersmith, London). This process was replicated at the beginning of each testing day, and once every ten samples thereafter, in accordance with the manufacturers guidelines.

Haemoglobin - prior to testing calibration was checked daily using a control cuvette standard (HemoCue®, Ängelholm, Sweden). In addition, samples were taken in duplicate, and analysed immediately after checking the optical eye for air. Also, this procedure complied with the manufacturer's guidelines.

Power output - The cycle ergometer was calibrated daily, by following the computer prompted procedures (Coleman, 1996).

Statistical Procedures

All statistical procedures were carried out via the statistical computer package SPSS (SPSS Version 11, SPSS Inc., Chicago, IL). Firstly, a one-way ANOVA was performed to check for significant F values. Next, pair-wise comparisons were made using a Tukeys Honest Significant Difference ($Tukey^{HSD}$), Post-Hoc test. Significance was accepted at the $P<0.05$ level. Unless otherwise stated, all results are presented as means ± SD.

Terminology

Throughout the study, Peak Power Output (PPO) and Mean Power Output (MPO) values were obtained for each sprint. PPO refers to the highest power value obtained during 1-s of each 6-s sprint. Whereas, MPO refers to the average power obtained during each sprint, as calculated by the sum of the PPO achieved in each 1-s integral of each sprint divided by 6 ($\sum$ power outputs/6). In addition, the fatigue rate was calculated as the difference between the mean of the highest two peak powers and the lowest two peak power values, in each sprint. This was then expressed as a percentage of the highest two peak power values.

Results

Power Output: The PPO and MPO recorded during each of the eight 6-s sprints are shown in figure 1. The time taken to achieve PPO in all sprints was within the first 2-s of exercise. The high peak power generated during the first sprint (874.9 ± 175.6 W) further substantiates the maximal nature of the exercise test. Maximal PPO did not occur in the first sprint as one would suspect, instead it occurred during the third sprint (882.6 ± 172.9 W). Despite this factor, following sprint three there was a trend for the power output to decline but only to ~80% of that in the first sprint by sprint eight (715.9 ± 147.1 W). In summary, although PPO declined with successive bouts of maximal sprinting, none of the differences observed were significant at the $P < 0.05$ level.

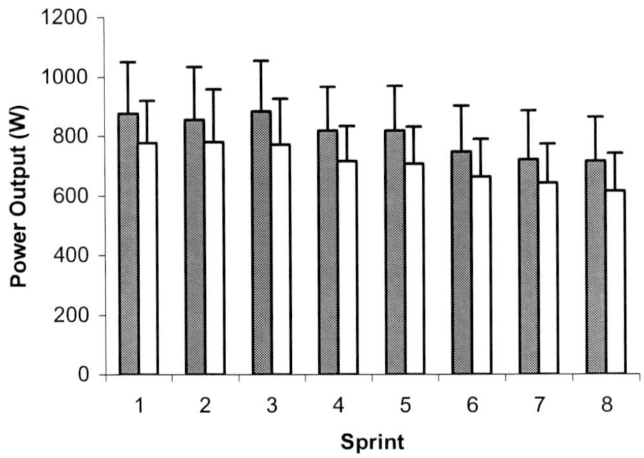

Figure 1. Peak power output (shaded columns) and mean power output (open columns) observed for eight 6-second sprints separated by 60-seconds of seated rest. Values are mean ± SD; n = 8.

Blood Lactate: Following 15min of seated rest, blood lactate concentrations ($B_{[LA]}$) were marginal (0.67 ± 0.47 mmol/L), the first sprint resulted in a small increase in $B_{[LA]}$ (mean Δ Lactate 0.75mmol/L), but was insignificant ($P > 0.05$). $B_{[LA]}$ continued to increase with successive sprints and had increased significantly immediately after sprint three (mean Δ Lactate, 3.23 mmol/L, $P < 0.001$). The highest value was obtained immediately after the last sprint, as was expected (7.5 ± 1.8 mmol/L) (figure 2). This value was also significantly higher than the values obtained after sprints one, two and three ($P < 0.0001$) and significantly higher than the resting value.

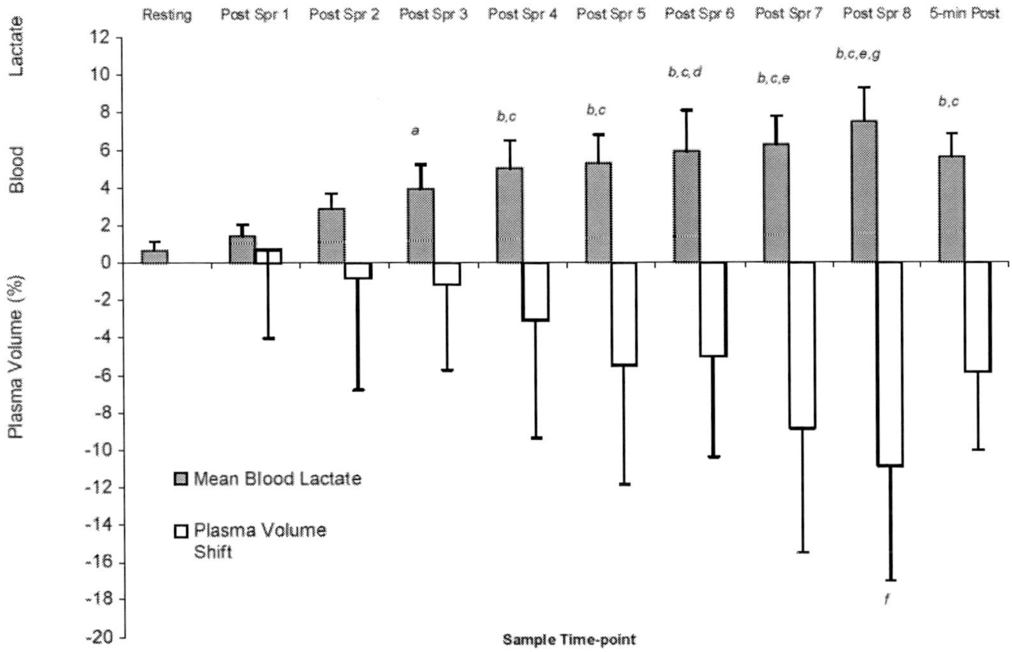

Figure 2. Blood lactate results and plasma volume changes observed for eight 6-second sprints separated by 60-seconds of seated rest. Values are Mean ± SD, n = 8. Resting, taken after 15min seated rest; Post Spr, taken immediately after sprint (corresponding number); [a], significantly different to resting at P< 0.001; [b], different to resting at P< 0.0001; [c], significantly different to post sprint 1 at p< 0.0001; [d], significantly different to Post Sprint 2 at p< 0.001; [e], significantly different to Post Sprint 2 at p<0.0001; [f], Significantly different to Post sprint 2 at p< 0.005 and [g], significantly different to Post Sprint 3 at p<0.0001.

Plasma Volume: The initial response was a subtle haemodilution (+0.69 ± 4.77 %,) following sprint one. To the author's knowledge this is the first time, such an observation has been made during this type of protocol. However although this may represent a response of the body's homeostatic mechanisms, due to the large measures of central tendency, the haemodilution could quite well be an artefact of the subjects differing training and/or hydration status. Nevertheless, on the whole there followed a gradual haemoconcentration after sprint two (Post sprint 2, -0.86 ± 5.94%), and approached a significant change from the resting value immediately after sprint eight (~9.5 % haemoconcentration, P < 0.05). In addition, post sprint eight there was a significant decrease in plasma volume compared to post sprint two (Δ Plasma volume –10%).

Heart rate: Heart rates were significantly elevated after each sprint, compared to the resting value (P < 0.0001). The peak heart rate occurred immediately after sprint seven (143 ± 16 bpm) and was double the value recorded at rest (69 ± 7 bpm). The author would like to acknowledge that the relatively sub-maximal heart rates, reflects the experimental methods more so than the actual nature of the test, since no values were obtained during the sprints, but were all taken during recovery. Recovery period heart rates are represented in figure 3.

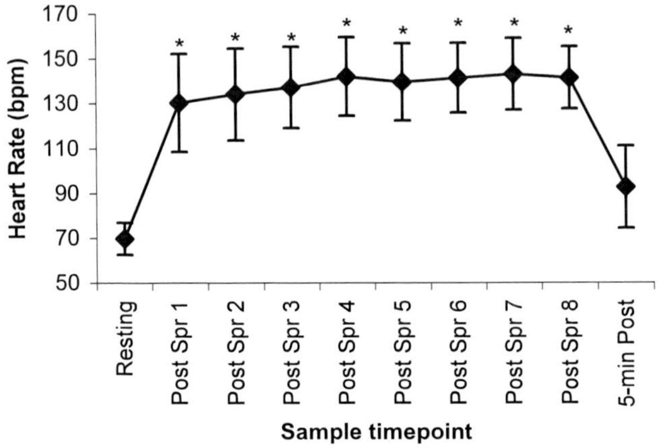

Figure 3. Mean heart rates observed for eight 6-second sprints separated by 60-seconds of seated rest. Values are Mean ± SD.; n = 8. * significantly different to resting value (P<0.0001).

Fatigue index (FI): FI values are presented in figure 4. The mean FI throughout all time-points was 16.7 ± 8.3%, the largest value's for FI occurred during sprints one and seven (19.1 ± 9.6% and 19.5 ± 15.4 %, respectively). However, there were no significant differences between any of the sprints (P > 0.05).

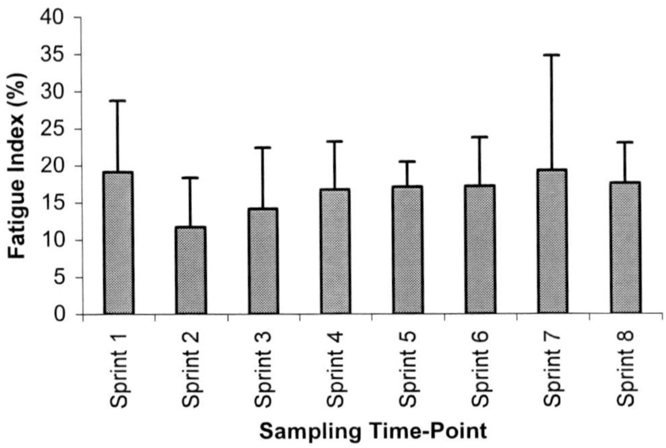

Figure 4. Mean Fatigue Index observed for eight 6-second sprints separated by 60-seconds of seated rest. Values are Mean ± SD.; n = 8.

Discussion

The main findings of this study were that 60-s of recovery from brief maximal exercise is sufficient to support the near complete recovery from short term intense exercise, presumably caused by the adequate replenishment of the anaerobic energy stores (ATP-PC). The significant increases in blood lactate during the experiment reveal that anaerobic glycolysis played a significant role in energy provision as exercise progressed.

Although all of the eight 6-s sprints were performed maximally, there were no significant reductions in power output throughout the duration of the test. These results conflict with those by Gaitanos et al (1993) who showed a marked decrease in both MPO and PPO after sprint 4 and reductions of 26.6% and 34.4% respectively by the final sprint (Sprint 10). This apparent disparity was most probably due to Gaitanos et al. using 30-s of recovery, compared to the present study's 60-s of recovery. Since it would make great 'metabolic sense' that the longer the recovery duration, the more adequately the ATP stores could be replenished. This is especially true given the known rapid initial phase of creatine phosphate recovery from intense exercise.

It has previously been shown that, ATP turnover is directly related to the force generated during dynamic exercise (Hultman et al. 1983). The forces produced in the present study were proportionally high (874.9 ±175.6 W), when compared to consecutive Wingate protocols (668 ± 99 W) (Backx et al. 2000) or sprinting on a non-motorised treadmill (534 ± 85 W) (Cheetham et al. 1986), but were less than those reported for a similar protocol (1253.3 ± 334.8 W) (Gaitanos et al. 1993). At such high power outputs, the oxidative mechanisms would invariably have been rate limited (Spriet et al. 2000) requiring an almost exclusive amount of energy to have been derived from the anaerobic sources, namely, the degradation of PCr in the creatine kinase reaction and in the glycolytic pathway via the conversion of glycogen to lactate. In addition, the interesting findings that neither PPO, MPO or fatigue index decreased with successive sprints, further substantiate the findings that 60-s was enough for the anaerobic energy systems to recover from repetitive bouts of all-out physical exertion.

Both the time-course and magnitude of change in blood lactate concentration has significant implications. Firstly, the manner in which the lactate concentration increased adds further credence to the work of Wooton and Williams (1983), supporting their findings that intermittent maximal exercise leads to an increasing demand on anaerobic glycolysis to maintain the rate of energy production. Secondly, since the blood lactate concentrations observed (7.5 ± 1.8 mmol/L) closely parallel those seen during a competitive rugby union match (7.2 mmol/l) (Deutsch et al. 1998) and those observed during a controlled field test that mimicked the activity patterns of soccer (7.0 ± 1.2 mmol/L) (Nicholas et al. 1999), there maybe usefulness in devising a laboratory based experimental procedure, as a method for the scientific study of high intensity interval training for field sports during repeat sprint activity. However, since this hypothesis was not part of the initial experimental aims, these findings would have to be elucidated by further research.

Previous studies have consistently shown that recovery from high-intensity exercise is dramatically improved in aerobically fit individuals (for comprehensive review, see Tomlin and Wenger 2001). Some of the proposed mechanisms for these training-induced adaptations include increased temperature regulation, better metabolism/utilisation of fuel substrates, and the selective recruitment and hypertrophy of both slow and fast-twitch muscle fibres (Tomlin and Wenger, 2001). Tomlin and Wenger suggest these adaptations primarily help by favouring the oxidation of lactate to pyruvate, which would provide a ready fuel, for aerobic metabolism and help normalise pH by consuming H^+. These adaptations at least in part, could help to explain the relatively large variances observed for a number of the results obtained. Moreover, they help explain why there was a marked increase in ventilation during the recovery periods (authors observation) and a pronounced elevation in heart rate throughout rest periods, as both homeostatic responses would seek to increase oxygen supply to the muscle, which would ultimately aid aerobic metabolism.

Factors known to influence plasma volume changes are both varied and multi-faceted (Kargotich et al. 1998). Despite the best efforts of the investigators to control for known confounding factors such as, heat stress, posture and hydration status, large variances still may have existed. This made it increasingly difficult to detect significant changes with the statistical methods employed. Nevertheless, from observations made during testing and subsequently reviewing the data we believe the plasma volume was critically influenced by increased perspiration, increased insensible fluid loss and an increased myocyte osmolarity caused by the muscles accruing lactate. This latter finding was based on the fluid losses in the plasma negatively mimicking the rise in blood lactate concentration.

In conclusion, the results of this research enable the researchers to reject the null hypotheses proposed. Blood lactate clearly accumulated with successive sprints. The findings of this study have far-reaching implications for the further understanding of intermittent exercise bouts. Also, if capillary blood samples are an indication of metabolism at the muscular level as a number of researchers suggest (Cheetham 1986) eight, 6-s maximal sprints interspersed with 60-s of static recovery, undoubtedly involves an appreciable contribution of the glycogenolytic/glycolytic anaerobic pathways to muscle ATP turnover.

References

Backx K., McNaughton, L., Crickmore L., Palmer G., and Carlisle, A. 2000. Effects of differing heat and humidity on the performance and recovery from multiple high intensity, intermittent exercise bouts. *Int. J. Sports Med.* 21: 400-405.

Cheetham, M., Boobis, L., Brook, S., and Williams, C. 1986. Human muscle metabolism during sprint running. *J. Appl. Physiol.* 61: 54-60.

Coleman S (1996) Corrected Wingate Anaerobic Test. Cranlea and Co, Sandpits Lane, Acacia Rd, Bournville, B'ham, U.K: 4-5.

Deutsch, M.U., Maw, G.J., Jenkins, D., and Reaburn, P. 1998. Heart rate, blood lactate and kinematic data of elite colts (under–19) rugby union players during competition. *J. Sports Sci.* 16: 561-570.

Dill, D.B., and Costill, D.L. 1971. Calculation of percentage changes in volumes of blood, plasma, and red cells in dehydration. *J. Appl. Physiol.* 37(2): 247-248.

Gaitanos, G.C., Williams, C., Boobis, L.H., and Brooks S. 1993. Human muscle metabolism during intermittent maximal exercise. *J. Appl. Physiol.* 75(2): 712-719.

Gastin, P.B. 2001. Energy system interaction and relative contribution during maximal exercise. *Sports Med.* 31(10): 725-741.

Hermansen, L., and Vaage, O. 1977. Lactate disappearance and glycogen synthesis in human muscle after maximal exercise. *Am. J. Physiol.* 233(5): 422-429.

Heyward, V.H. 1991. Advanced fitness assessment and exercise prescription (3rd Edition). Human Kinetics, Champaign, IL., p.147.

Hultman E, Chasiotis D, Sjöholm H. 1983 Energy metabolism in muscle. *Prog. Clin. Biol. Res.* 136:257-72.

Jackson AS. 1984 Research design and analysis of data procedures for predicting body density. *Med. Sci. Sports Exerc.* 16(6):616-22.

Jacobs, I. 1986. Blood lactate: Implications for training and sports performance. *Sports Med.* 3: 22-37.

Kargotich, S., Goodman, C., Keast, D., and Morton, A.R. 1998. The influence of exercise-induced plasma volume changes on the interpretation of biochemical parameters used for monitoring exercise, training and sport. *Sports Med.* 26(2): 101-117.

Lakomy, H. 1986. Measurement of work and power output using friction-loaded Cycle ergometers. *Ergonomics.* 29: 509-517.

Lohman, T.G., Roche, A.F., and Martorell, R. 1988. Anthropometric standardisation reference manual. *Human Kinetics,* Champaign, IL., p.55-69.

McCartney, N., Spriet, L.L., Heigenhauser, G.J.F., Kowalchuk, J.M., Sutton, J.R., and Jones, N.L. 1986. Muscle power and metabolism in maximal intermittent exercise. *J. Appl. Physiol.* 60(4): 1164-1169.

Nicholas CW, Tsintzas K, Boobis L, Williams C. 1999 Carbohydrate-electrolyte ingestion during intermittent high-intensity running. *Med. Sci. Sports Exerc.* 31(9):1280-6.

Rodas, G., Ventura, J.L., Cadefau, J.A., Cusso, R., and Parra, J. 2000. A short training programme for the rapid improvement of both aerobic and anaerobic metabolism. *Eur. J. Appl. Physiol.* 82: 480-486.

Saltin, B., and Essen, B. 1971. Muscle glycogen, lactate, ATP and CP in intermittent exercise. *In* Muscle metabolism during exercise: Advances in Experimental Medicine and Biology. Vol. Ll. E*dited by* B. Pernow and B. Saltin. Plenum, New York, pp. 419-424.

Smith, J.C., Hill, D.W. 1991. Contribution of energy systems during a Wingate power test. *Br. J. Sp. Med.* 25(4): 196-199.

Spriet, L.L., Howlett, A.R., and Heigenhauser, J.F.G. 2000. An enzymatic approach to lactate production in human skeletal muscle during exercise. *Med. Sci. Sports Exerc.* 32(4): 756-763.

Tomlin, L.D., and Wenger, A.H. 2001. The relationship between aerobic fitness and recovery from high intensity intermittent exercise. *Sports Med.* 31(1): 1-11.

Welsh, R.S., Davis, J.M., Burke, J.R., and Williams H.G. 2002. Carbohydrates and physical/mental performance during intermittent exercise to fatigue. *Med. Sci. Sports Exerc.* 34(4): 723-731.

In: Glycolysis: Regulation, Processes and Diseases
Editor: Paul N. Lithaw
ISBN: 978-1-60741-103-1
© 2009 Nova Science Publishers, Inc.

Chapter IX

Mathematical Modeling as a Tool for Decoding the Control of Metabolic Pathways

Eberhard Voit*

Integrative BioSystems Institute and The Wallace H. Coulter Department of Biomedical Engineering, Georgia Institute of Technology, Atlanta, Georgia 30332-0535, USA

Abstract

Glycolysis is probably the best understood biochemical pathway. It has been subjected to about every imaginable type of investigation, from phenomenological observations to detailed analyses of its components with methods of enzyme kinetics and *in vivo* nuclear magnetic resonance. In many ways, the gradual increase in information and knowledge associated with the glycolytic pathway can be seen as a representative of the growing body of insights into metabolism in general. The development of mathematical and computational models of glycolysis has mirrored the experimental exploration, although with substantial time delay. Indeed, models of glycolysis can be seen as sentinels of important phases of metabolic model creation, including the choices of model types at different times and the purposes for creating these models. Early models were designed as proof of concept that mathematical equations were capable of capturing biological observations. Some of these early, simple models eventually grew into comprehensive descriptions of glycolysis in different contexts and with species dependent variations, allowing detailed simulations of what-if scenarios. Other models stayed intentionally simple in order to allow the extraction and rigorous mathematical analysis of the essence of the pathway, for instance, with respect to oscillations. Some of the models were used for optimization within a context of metabolic engineering, others as means of explaining non-intuitive features of pathway control and regulation. This chapter reviews some of these developments and demonstrates how they have been

* Contact: Georgia Institute of Technology, 313 Ferst Drive, Suite 4103, Atlanta, Georgia 30332-0535, USA; email: eberhard.voit@bme.gatech.edu

leading to the present-day frontiers of discovering design and operating principles and to guidance for the creation of pathway systems in the new field of synthetic biology.

A Brief History of Models of Glycolysis

Glycolysis is without doubt the best studied and best understood metabolic pathway. It is also a paradigm for how biochemistry has changed throughout its existence. Interestingly, we find the same role of glycolysis as a sentinel and representative of mathematical modeling in biochemistry and metabolism. To see the parallels, it is necessary to go back in time, review developments on the experimental and the modeling fronts, determine where we stand at this point in time, and speculate where we might go from here.

Five millennia ago, food engineers in Egypt discovered that they could combine bread and beer production by using beer mash and foam as starters for leavening bread (Enfors 2001). Whether this discovery was by chance or ingenuity will probably remain a secret of history, but it is safe to assume that these early bakers and brewers did not know about yeast, lactic acid bacteria, glycolysis or the molecular underpinnings of fermentation. Even without a solid scientific background, they managed to control the relevant biochemical processes by determining just the right amounts of flour, water and starter, and by carefully controlling temperature and other conditions. Fast forward almost five thousand years to 1857, when Louis Pasteur made the connection between fermentation and yeast. As he famously wrote: "It is my opinion that alcoholic fermentation never occurs without the simultaneous organization, development and multiplication of cells." Fifty years later, Eduard Buchner received the Nobel Prize in Chemistry for determining that yeast secreted a substance he termed zymase, which he considered responsible for the fermentation process. In the first half of the 20th Century, Gustav Georg Embden, Otto Fritz Meyerhof, Luis Federico Leloir and many others elaborated the steps in the generation of lactic acid from glucose in what became known as the "Embden-Meyerhof pathway." By now, we believe to have a relative comprehensive and detailed picture of glycolysis, its central role in ATP generation, its side reactions, and the various fates of pyruvate, which may be converted into a number of other metabolites, such as citric acid with all its derivatives, acetate and ethanol. We also realize that this nearly ubiquitous sequence may show numerous variations in many of its components. Most organisms use ATP for powering the first phosphorylation, others phosphoenolpyruvate (PEP). Some organisms generate ethanol, others lactate; some organisms use glycolysis for aerobic, others for anaerobic respiration. While we are relatively certain about the metabolic conversions, the regulation of the pathway, in concert of genes and metabolic modulators, is often still not quite clear (Teusink, Passarge et al. 2000). As just one example, the up-regulation of glycolytic genes under stress conditions may exhibit patterns that are far from being intuitive (Voit and Radivoyevitch 2000).

The development of mathematical models of glycolysis has been paralleling both the experimental findings (although with a long time delay) and the growth of the broader field of metabolic pathway analysis and biochemical systems modeling. Arguably the earliest "algebraic equation" in biochemistry was formulated in the late 18th century by the famous oxygen and respiration scientist Antoine Lavoisier. In collaboration with the great mathematician Pierre Simon Laplace he posited "must [juice] of grapes = carbonic acid + alcohol," explaining that "by successively supposing each of the elements in this equation unknown, we can calculate their values in succession and thus verify our experiments by calculation and our calculations by experiments reciprocally" ((Leicester 1974): p.139).

Two-hundred years later, intrigued by the observation that yeast cells may show metabolic oscillations (*e.g.*, (Pye 1966; Pye and Chance 1966; Pye 1969)), some of the early mathematical modelers began developing kinetic and dynamic models of glycolysis that were able to mimic these oscillations (*e.g.*, (Higgins 1964; Sel'kov 1968; Mochan and Pye 1973). While mathematical formulations of enzymatic processes had been known since Henri, Hill, Michaelis and Menten (Henri 1903; Hill 1910; Michaelis and Menten 1913), and many others (see (Schulz 1994)), a significant purpose of the new breed of models was to demonstrate the ability to translate entire chains or even networks of enzymatic reactions into dynamic mathematical systems of equations (*e.g.*, (Heinrich and Rapoport 1973; Rapoport and Heinrich 1975; Rapoport, Heinrich et al. 1976)). This transition required the jump from explicit function descriptions of individual enzyme-catalyzed steps to systems of nonlinear ordinary equations. Indeed, the parameter values in these dynamic models were reasonable, and the model responses were often consistent with experimental findings. Thus, it had been proven in principle that it was possible to translate metabolic pathway systems into mathematics. Supported by the arrival and accessibility of computers, some investigators were convinced that these representations could be expanded almost limitlessly to models of arbitrarily large metabolic systems (Garfinkel 1968; Garfinkel, Frenkel et al. 1968; Gavalas 1968; Garfinkel 1980).

Even at the time of these early successes, the first warnings were issued. Savageau (Savageau 1970) pointed out that systems consisting of rate functions in the tradition of Michaelis and Menten would require infeasibly large numbers of biochemical experiments for parameter estimation and subsequent simulation studies. Furthermore, such an approach, because of its particular and essential nonlinearities, would preclude mathematical analyses and only allow simulation studies, which could never reach the rigor of analytical mathematics. In fact, Savageau suggested that there were not even the tools, let alone a theory, for objectively comparing and assessing different pathway structures and their features (Savageau 1972). As a consequence, he concluded that only effective approximations would provide suitable alternatives and proposed the modeling framework of Biochemical Systems Theory (BST; (Savageau 1969; Savageau 1969)), which by now has been the basis for hundreds of articles and several books (Savageau 1976; Voit 1991; Voit 2000; Torres and Voit 2002; Shiraishi 2007; Goel 2008). Similar to Savageau's concerns, Heinrich and Rapoport (Heinrich and Rapoport 1974) discussed the limitations of simply using rational rate laws in ever growing metabolic models. They pointed to the intrinsic difficulties in differentiating between important and unimportant effects and components and in establishing cause-effect relationships in complex networks, along with the need of data and assumptions that were not always easy to obtain. As an alternative they suggested Metabolic Control Analysis (Kacser and Burns 1973; Heinrich and Rapoport 1974), which also resulted in numerous articles, book chapters and books (*e.g.*, (Cornish-Bowden and Cárdenas 1990; Fell 1997)).

The opportunities afforded by computers on one hand and the warnings issued about pure computer models on the other formed the basis from which further work in computational biochemistry derived, and this work is again easily demonstrated in the context of glycolysis. One stream of modeling research pursued the path of ever increasing complexity and detail, resulting in very comprehensive models of glycolysis (*e.g.*, (Rizzi, Baltes et al. 1997; Teusink, Passarge et al. 2000; Hynne, Danø et al. 2001)). Other work connected glycolysis to adjacent pathways in order to understand cross-talk between pathway systems (*e.g.*, (Mulquiney and Kuchel 1999; Lambeth and Kushmerick 2002)). These large-scale models have reached a degree of sophistication that permits realistic simulations studies and support the exploration of what-if scenarios.

In contrast to the search for comprehensive coverage, other models of glycolysis targeted the structural and regulatory essence of pathways. For instance, they aimed to identify the pathway features that were minimally needed to explain the observed oscillations in the glycolytic pathway (see above and (Richter, Betz et al. 1975; Goldbeter 1996; Hynne, Danø et al. 2001; Westermark and Lansner 2003)).

A different line of research targeted practical applications in metabolic engineering, for instance with the goal of manipulating microorganisms in such a fashion that they produce more of a bulk product, such as ethanol or citric acid, or a pure compound for the pharmaceutical or food industry (Stephanopoulos, Aristidou et al. 1998). In the past, this goal had been pursued primarily with the experimental method of random mutagenesis and selection, in which the microorganisms were subjected to adverse conditions under which they naturally mutated. The mutants best matching the design goals were then selected. Because of the random nature of this process, improvements in strains tend to become successively more difficult to obtain. An alternative is the construction of a whole cell model, or at least a model of metabolism, describing the pathways of interest and allowing the screening for effective manipulations. For example, Curto and collaborators (Cascante, Curto et al. 1995; Curto, Sorribas et al. 1995; Sorribas, Curto et al. 1995) developed a dynamic model of glycolysis, which was later used for optimization studies in the context of metabolic engineering (Crawford and Blum 1993; Torres, Voit et al. 1997; Torres and Voit 2002; Schwender, Ohlrogge et al. 2003; Polisetty, Voit et al. 2008). Optimization approaches were also used to explain the variations in glycolysis throughout evolutionary mechanisms (Heinrich, Montero et al. 2004).

Finally, mathematical models of glycolysis and other metabolic pathways can be used for the discovery of design and operating principles (*e.g.*, (Savageau 1976; Irvine and Savageau 1985; Savageau 1985; Hlavacek and Savageau 1997; Alves and Savageau 2000; Schwacke and Voit 2004; Voit 2003)). The questions here are of a more general type, namely, one asks: why is the pathway organized in the fashion we observe and not in a different fashion? For instance, why is negative feedback in a linear pathway essentially always exerted by the end product onto the first step? Why does FBP feedforward-activate pyruvate kinase? These classes of questions are very difficult to answer with the means of an experimental laboratory, because it is difficult to generate a mutant in which, for instance, FBP does not affect pyruvate kinase, but is otherwise identical. By contrast, these questions of natural design are very much amenable to rigorous mathematical analysis. According to the method of controlled mathematical comparisons, which was developed specifically for the discovery of design principles, the analysis is executed simultaneously with two models: one representing the observed pathway and the other a hypothetically reasonable alternative. In the case of feedback inhibition, the first model would contain inhibition of the first step by the end product, while the alternative model would be exactly the same, but without the feedback or with a different type of feedback. Side-by-side comparisons of the two models, for instance with respect to stability, robustness, or response time, then allow objective assessments of the advantages and drawbacks of the observed design, in comparison to an alternative.

As two specific examples in the context of glycolysis, we analyzed the fermentation pathway in *Saccharomyces cerevisiae* under heat stress conditions (Voit and Radivoyevitch 2000) and the same pathway in *Lactococcus lactis* under conditions of aerobic lactate production (Voit, Almeida et al. 2006). The former allowed us to rationalize non-intuitive gene regulation pattern, while the latter permitted us to characterize the specific role of the very rare feedforward activation of pyruvate kinase by fructose-1,6-bisphosphate and the simultaneous inhibition of the same step by inorganic phosphate with respect to the proper functioning of the phosphoenolpyruvate: carbohydrate phosphotransferase system (PEP:PTS). The mathematical modeling methods also rendered it possible to characterize the relative importance of the activating and inhibiting control signals (Voit, Neves et al. 2006).

The Future

Richard Feynman, one of the legendary physicists of our times, once made the now famous comment: "What I cannot create I do not understand." In the study of glycolysis, and of metabolic pathways in general, the act of creating can be pursued in three ways. The first is the manipulation of altered pathway features in metabolic engineering. In most cases, these alterations do not change the pathway of interest structurally but target the flux through the system, which may be affected by changes in the organism's genetic make-up that in turn lead to altered enzyme activities or block less desired pathways that use the same substrates. The introduction of alterations has often been attempted with random mutagenesis, followed by selection, or through the activation of relevant genes by means of plasmids or the insertion of promoters.

The second option of "creation" is the construction of mathematical models. These are by their nature different in character from actual molecules and enzyme catalyzed processes, but they allow us to analyze different types of features of a pathway. Not only is it easy to perform what-if simulations, mathematical models are also ideally suited for all kinds of manipulations and optimizations, even under stringent sets of constraints. They render it feasible to classify cellular responses into what is feasible, probable, and impossible.

Finally, mathematical models offer us the chance of discovering the natural design and operating principles that govern pathways and pathway systems. This discovery is the basis for the third option of creation, which is at the core of synthetic biology. It is possible with today's methods to design small gene circuits that lead to the predictable production of target proteins (*e.g.*, (Guet, Elowitz et al. 2002; Atkinson, Savageau et al. 2003; Rosenfeld, Young et al. 2005)). The future will allow us to create much more complex systems. Steering this creation will require that we fully understand not only the components and the individual processes, but also their integrative behavior. This understanding, in turn, will almost certainly require the support from mathematical and computational models.

Acknowledgments

This work was supported in part by a Molecular and Cellular Biosciences Grant (MCB-0517135; E.O. Voit, PI) from the National Science Foundation, and an endowment from the Georgia Research Alliance. Any opinions, findings, and conclusions or recommendations expressed in this material are those of the authors and do not necessarily reflect the views of the sponsoring institutions.

References

Alves, R. and M. A. Savageau (2000). "Comparing systemic properties of ensembles of biological networks by graphical and statistical methods." *Bioinformatics*. 16(6): 527-33.

Atkinson, M. R., M. A. Savageau, et al. (2003). "Development of genetic circuitry exhibiting toggle switch or oscillatory behavior in Escherichia coli." *Cell.* 113(5): 597-607.

Cascante, M., R. Curto, et al. (1995). "Comparative characterization of the fermentation pathway of Saccharomyces cerevisiae using biochemical systems theory and metabolic control analysis. Steady-state analysis. ." *Math. Biosc.* 130(51-69).

Cornish-Bowden, A. and M. L. Cárdenas, Eds. (1990). Proceedings of the Advanced NATO Research Workshop on Control in Metabolic Systems. New York, NY, Plenum Press.

Crawford, J. M. and J. J. Blum (1993). "Quantitative analysis of flux along the gluconeogenic, glycolytic and pentose phosphate pathways under reducing conditions in hepatocytes isolated from fed rats." *Biochem. J.* 212(3): 585-598.

Curto, R., A. Sorribas, et al. (1995). "Comparative characterization of the fermentation pathway of Saccharomyces cerevisiae using biochemical systems theory and metabolic control analysis. Model definition and nomenclature. ." *Math. Biosc.* 130: 25-50.

Enfors, S. O. (2001). Baker's yeast. Basic biotechnology. C. Ratledge and B. Kristiansen. Cambridge, U.K. ; New York, NY, Cambridge University Press.

Fell, D. (1997). Understanding the control of metabolism. London, U.K., Portland Press.

Garfinkel, D. (1968). "The role of computer simulation in biochemistry." *Comp. & Biomed. Res.* 2(1): 31-44.

Garfinkel, D. (1980). " Computer modeling, complex biological systems, and their simplifications. ." *Amer. J. Phys.* 239(1): R1-R6.

Garfinkel, D., R. A. Frenkel, et al. (1968). "Simulation of the detailed regulation of glycolysis in a heart supernatant preparation." *Comp. & Biomed. Res.* 2(1): 68-91.

Gavalas, G. R. (1968). *Nonlinear Differential Equations of Chemically Reacting Systems. .* Berlin, Springer-Verlag.

Goel, G. (2008). *Reconstructing Biochemical Systems. Systems Modeling and Analysis Tools for Decoding Biological Designs.* Saarbrücken, Germany, VDM Verlag Dr. Müller.

Goldbeter, A. (1996). *Biochemical Oscillations and Cellular Rhythms.* Cambridge, Cambridge University Press.

Guet, C. C., M. B. Elowitz, et al. (2002). "Combinatorial synthesis of genetic networks." *Science.* 296(5572): 1466-70.

Heinrich, R., F. Montero, et al. (2004). "Theoretical Approaches to the Evolutionary Optimization of Glycolysis. Thermodynamic and Kinetic Constraints." *Eur. J. Biochem.* 243(1-2): 191-201.

Heinrich, R. and T. A. Rapoport (1973). "Linear theory of enzymatic chains; its application for the analysis of the crossover theorem and of the glycolysis of human erythrocytes." *Acta Biol. Med. Ger.* 31(4): 479-494.

Heinrich, R. and T. A. Rapoport (1974). "A linear steady-state treatment of enzymatic chains: General properties, control and effector strength " *Eur. J. Biochem.* 42: 89-95.

Henri, M. V. (1903). *Lois générales de l'action des diastases.* Paris, Hermann.

Higgins, J. (1964). "A Chemical Mechanism for Oscillation of Glycolytic Intermediates in Yeast Cells." *Proc. Natl. Acad. Sci. U. S. A.* 51: 989-94.

Hill, A. V. (1910). "Possible effects of the aggregation of the molecules of haemoglobin on its dissociation curves." *J. Physiol.* 40: iv-viii.

Hlavacek, W. S. and M. A. Savageau (1997). "Completely uncoupled and perfectly coupled gene expression in repressible systems." *J. Mol. Biol.* 266(3): 538-58.

Hynne, F., S. Dano, et al. (2001). "Full-scale model of glycolysis in Saccharomyces cerevisiae." *Biophys. Chem.* 94(1-2): 121-163.

Irvine, D. H. and M. A. Savageau (1985). "Network regulation of the immune response: alternative control points for suppressor modulation of effector lymphocytes." *J. Immunol.* 134(4): 2100-16.

Kacser, H. and J. A. Burns (1973). "The control of flux." Symp Soc Exp Biol 27: 65-104.

Lambeth, M. J. and M. J. Kushmerick (2002). "A computational model for glycogenolysis in skeletal muscle." *Ann. Biomed. Eng.* 30(6): 808-27.

Leicester, H. M. (1974). *Development of Biochemical Concepts from Ancient to Modern Times.* . Cambridge, Mass, Harvard University Press.

Michaelis, L. and M. L. Menten (1913). "Die Kinetic der Invertinwirkung." *Biochem. Zeitschrift.* 49: 333-369.

Mochan, E. and E. K. Pye (1973). "Respiratory oscillations in adapting yeast cultures." *Nat. New Biol.* 242(119): 177-9.

Mulquiney, P. J. and P. W. Kuchel (1999). "Model of 2,3-bisphosphoglycerate metabolism in the human erythrocyte based on detailed enzyme kinetic equations: equations and parameter refinement." *Biochem. J.* 342: 581-596.

Polisetty, P. K., E. O. Voit, et al. (2008). "Yield Optimization of Regulated Metabolic Systems Using Deterministic Methods." *Biotech. Bioengin.* 99(5): 1154-1169.

Pye, E. K. (1966). "Metabolic control phenomena associated with oscillatory reactions." *Studia Biophysica (Berlin).* 1:a75-a78.

Pye, E. K. (1969). "Biochemical mechanisms underlying the metabolic oscillations in yeast." *Can. J. Bot.* 47: 271-285.

Pye, K. and B. Chance (1966). "Sustained sinusoidal oscillations of reduced pyridine nucleotide in a cell-free extract of Saccharomyces carlsbergensis." *Proc. Natl. Acad. Sci. U. S. A.* 55(4): 888-94.

Rapoport, T. A. and R. Heinrich (1975). "Mathematical analysis of multienzyme systems. I. Modelling of the glycolysis of human erythrocytes." *Biosystems.* 7(1): 120-129.

Rapoport, T. A., R. Heinrich, et al. (1976). "The regulatory principles of glycolysis in erythrocytes in vivo and in vitro. A minimal comprehensive model describing steady states, quasi-steady states and time-dependent processes." *Biochem. J.* 154: 449-469.

Richter, O., A. Betz, et al. (1975). "The response of oscillating glycolysis to perturbations in the NADH/NAD system: a comparison between experiments and a computer model." *Biosystems.* 7: 137-146.

Rizzi, M., M. Baltes, et al. (1997). "In Vivo Analysis of Metabolic Dynamics in Saccharomyces cerevisiae:II. Mathematical Model." *Biotechnology and Bioengineering.* 55: 592-608.

Rosenfeld, N., J. W. Young, et al. (2005). "Gene regulation at the single-cell level." *Science.* 307(5717): 1962-5.

Savageau, M. A. (1969). "Biochemical systems analysis. I. Some mathematical properties of the rate law for the component enzymatic reactions." *J. Theor. Biol.* 25(3): 365-9.

Savageau, M. A. (1969). "Biochemical systems analysis. II. The steady-state solutions for an n-pool system using a power-law approximation." *J. Theor. Biol.* 25(3): 370-9.

Savageau, M. A. (1970). "Biochemical systems analysis. 3. Dynamic solutions using a power-law approximation." *J. Theor. Biol.* 26(2): 215-26.

Savageau, M. A. (1972). "The behavior of intact biochemical control systems." *Curr. Topics Cell. Regulation.* 6: 63-129.

Savageau, M. A. (1976). *Biochemical systems analysis : a study of function and design in molecular biology.* Reading, Mass., Addison-Wesley Pub. Co. Advanced Book Program.

Savageau, M. A. (1985). "A theory of alternative designs for biochemical control systems." *Biomed. Biochim. Acta.* 44(6): 875-80.

Schulz, A. R. (1994). *Enzyme kinetics : from diastase to multi-enzyme systems.* Cambridge ; New York, Cambridge University Press.

Schwacke, J. H. and E. O. Voit (2003). "Improved methods for the mathematically controlled comparison of biochemical systems." *BMC Theoretical Biology and Medical Modelling* **1**:1, 2004.

Schwender, J., J. B. Ohlrogge, et al. (2003). "A flux model of glycolysis and the oxidative pentosephosphate pathway in developing Brassica napus embryos." *J. Biol. Chem.* 278(32): 29442-53.

Sel'kov, E. E. (1968). "Self-oscillation in glycolysis. A simple kinetic model." *Eur. J. Biochem.* 4: 79-86.

Shiraishi, F. (2007). *BST: Biochemical Systems Theory. In Japanese.* Tokyo, Sangyo-tosho.

Sorribas, A., R. Curto, et al. (1995). "Comparative characterization of the fermentation pathway of Saccharomyces cerevisiae using biochemical systems theory and metabolic control analysis. Model validation and dynamic behavior." *Math. Biosc.* 130: 71-84.

Stephanopoulos, G. N., A. A. Aristidou, et al. (1998). *Metabolic Engineering. Principles and Methodologies.* San Diego, CA, Academic Press.

Teusink, B., J. Passarge, et al. (2000). "Can yeast glycolysis be understood in terms of in vitro kinetics of the constituent enzymes? Testing biochemistry." Eur. J. Biochem. 267(17): 5313-29.

Torres, N. V. and E. O. Voit (2002). *Pathway analysis and optimization in metabolic engineering.* New York, Cambridge University Press.

Torres, N. V., E. O. Voit, et al. (1997). "An indirect optimization method for biochemical systems: Description of method and application to the maximization of the rate of ethanol, glycerol, and carbohydrate production in Saccharomyces cerevisiae." *Biotech. Bioeng.* 55(5): 758-772.

Voit, E. O. (1991). *Canonical nonlinear modeling : S-system approach to understanding complexity.* New York, Van Nostrand Reinhold.

Voit, E. O. (2000). Computational analysis of biochemical systems : a practical guide for biochemists and molecular biologists. New York, Cambridge University Press.

Voit, E. O. (2003). "Design principles and operating principles: the yin and yang of optimal functioning." *Math. Biosci.* 182(1): 81-92.

Voit, E. O., J. S. Almeida, et al. (2006). "Regulation of Glycolysis in Lactococcus lactis: An Unfinished Systems Biological Case Study." *IEE Proc. Systems Biol.* 153: 286-298.

Voit, E. O., A. R. Neves, et al. (2006). "The Intricate Side of Systems Biology." *PNAS U.S.A.* 103(25): 9452-9457.

Voit, E. O. and T. Radivoyevitch (2000). "Biochemical systems analysis of genome-wide expression data." *Bioinformatics.* 16(11): 1023-1037.

Westermark, P. O. and A. Lansner (2003). "A Model of Phosphofructokinase and Glycolytic Oscillations in the Pancreatic ß-cell " *Biophys. J.* 85: 126-139.

In: Glycolysis: Regulation, Processes and Diseases
Editor: Paul N. Lithaw
ISBN: 978-1-60741-103-1
© 2009 Nova Science Publishers, Inc.

Chapter X

Influencing Metabolism during Critical Illness – Potential Novel Strategies

***NP Juffermans*[1,2]*, H Aslami*[1] *and MJ Schultz*[1,2,3]**
Academic Medical Centre, Amsterdam, the Netherlands
[1] Laboratory of Experimental Intensive Care and Anesthesiology (L.E.I.C.A.)
[2] Department of Intensive Care Medicine
[3] HERMES Critical Care Group, Amsterdam, the Netherlands

Abstract

Induced hypothermia after cardiopulmonary resuscitation ameliorates neurological outcome and is currently considered standard of care in clinical practice. An increasing amount of reports indicate that induced hypothermia is also beneficial in other conditions of hypoxia–induced organ injury, including intestinal ischemia–reperfusion injury and acute lung injury. Hydrogen sulphide, which inhibits oxidative phosphorylation, has been used to induce a suspended animation–like state in several rodent models, resulting in hypothermia and a reduction in metabolic rate. Hydrogen sulphide has been found to be protective against ischemia–reperfusion induced organ injury, including gut ischemia and acute lung injury.

In this manuscript, we speculate on the potential therapeutic effects of reducing metabolism in critically ill patients. In these patients, an exaggerated inflammatory response is common, which often results in multiple organ injury. Inducing a hypometabolic state during critical illness may limit organ injury by reducing oxygen consumption, a novel approach in the treatment of critically ill patients. Mitochondrial dysfunction during critical illness is described and the potential therapeutic possibilities of influencing metabolism during critical illness is discussed. Methods of inducing hypothermia and of inducing a suspended animation–like state with the use of hydrogen sulphide are described.

Introduction

In critical illness, a systemic hypermetabolic response is a common clinical entity. Sepsis is the exaggerated systemic inflammatory response to infection, characterized by hypotension as a result of vasodilatation, endothelial damage and microvascular dysfunction, ultimately resulting in impaired tissue oxygenation and organ injury [1]. Another vasodilatory shock state is the systemic inflammatory response syndrome (SIRS), which can occur as a reaction to a variety of non–infectious insults, including severe trauma, cardiothoracic surgery and ischemia–reperfusion injury. The dysregulated host inflammatory response in sepsis and SIRS can result, when untreated, in the multiple organ dysfunction syndrome (MODS). The development of MODS, including acute lung injury, or its more severe form acute respiratory distress syndrome, and acute kidney injury contribute strongly to morbidity and mortality in the critically ill [2].

The inflammatory response seen in MODS, requires an acceleration of glycolytic adenosine triphosphate (ATP)–supply to maintain the heightened level of activity and to prevent ATP levels from falling below threshold level, the latter which would compromise normal cell metabolism and trigger apoptotic cell death. Treatment of MODS traditionally consists of supportive care, ensuring adequate tissue perfusion and oxygenation to meet the high metabolic demands of severe inflammation. In this commentary, the potential beneficial effects of reducing metabolism in critically ill patients are discussed. Inducing a hypometabolic state may limit organ injury by restoring the dysbalance between oxygen demand and consumption, a novel therapeutic approach in critically ill patients.

Mitochondrial Dysfunction in Multiple Organ Failure

Etiology of Multiple Organ Failure in Critical Illness

The etiology of multiple organ dysfunction in critical illness is complex. A deficiency in tissue oxygen delivery has been ascribed to the development of organ failure. Both a failing cardiac output in relation to oxygen demand as well as shunting in the microcirculation may contribute to an inadequate organ perfusion and oxygenation. Shunting of the microcirculation is thought to result from increased arterial–venous flow through anatomical anastomoses, altered heterogeneity of the microvascular architecture and possibly the inability of hemoglobin to off–load oxygen fast enough to the tissues as it passes through the microcirculation [3]. However, despite apparent sufficient oxygen delivery, signs of hypoxia and/or metabolic dysfunction have been found to persist. Rather then caused by microcirculatory hypoxia, the tissue distress seen in MODS may be caused by disturbances in cellular metabolic pathways. The finding of an increased tissue oxygen tension in the presence of metabolic acidosis in sepsis suggests that oxygen is available at the cellular level and that the predominant defect may be a decreased use of oxygen in the mitochondria [4, 5].

Mitochondrial Function in Health

Under aerobic conditions, the mitochondrial oxidative phosphorylation system generates ATP to supply organs and tissues with energy needed for cellular function. The oxidation of glucose or lipids through the citric acid cycle, donates electrons to complexes I and II in the respiratory chain complex, which then flow through complexes III and IV. Protons are pumped out of the mitochondrial matrix into the inter–membrane space, resulting in the generation of an electrochemical proton gradient, which is used to drive ATP synthesis. Cytochrome c oxidase, the terminal enzyme of the respiratory chain complex, reduces oxygen to water, thereby consuming oxygen.

Mitochondrial Dysfunction in MODS

Mitochondrial abnormalities have been found in models of sepsis, reporting loss of structural integrity of mitochondrial membranes and swelling [6, 7]. Studies on mitochondrial function in models of sepsis have yielded variable results. Both an increase and a decrease in mitochondrial respiration have been reported (reviewed in [8]). These conflicting results have been contributed to the use of different species or organs between models, as well as to differences in the degree of resuscitation. However, in long–term sepsis models, a decrease in mitochondrial function is a consistent finding. Preclinical findings of mitochondrial dysfunction include the cytotoxicity of proinflammatory mediators. Tumor necrosis factor alpha (TNF) and nitric oxide (NO), both produced in excess during sepsis, affect oxidative phosphorylation by inhibiting several respiratory enzymes in the electron transport chain, thereby inducing direct functional damage to the mitochondria [9]. In accordance, most laboratory models of sepsis, have shown a decrease in mitochondrial activity and ATP generation [10-12]. Components of the electron transport chain may be variably damaged, although inhibition of complex I is most consistently involved [13-15]. The clinical relevance of mitochondrial dysfunction was shown in patients with septic shock [14]. It was found that the skeletal muscle ATP–concentration, a marker of mitochondrial oxidative phosphorylation, is depleted in septic shock, together with structural changes in the mitochondria. This impaired mitochondrial structure and function was associated with worse outcome [14]. Thus, bio–energetic failure (i.e. an inability to utilize oxygen), may be a mechanism underlying MODS in the critically ill.

Hypothesis of Mitochondrial "Shut–Down" During Critical Illness

It has been proposed that mitochondrial energy alterations are part of the strategic defense [16]. The perceived failure of organs might instead be a potentially protective mechanism. Reduced cellular metabolism could increase the chances of survival of cells, and thus organs, in the face of an overwhelming insult. In this view, the modifications induced by sepsis should not be regarded solely as a failure of energy cell status, but as an integrated response. The decline in organ function may be triggered by a decrease in mitochondrial activity and oxidative phosphorylation, leading to reduced cellular metabolism, the process of which may be triggered by acute changes in levels of hormones and inflammatory mediators. The fact that organ dysfunction is reversible in survivors of MODS, suggests a window of opportunity in which strategies to improve mitochondrial function may be possible.

Influencing Metabolism During Critical Illness

Enhancing Oxygen Delivery

In critically ill patients with MODS, treatment traditionally consists of supportive care, ensuring that high metabolic demands of severe inflammation are met by maintaining adequate tissue perfusion and oxygenation. Early correction of tissue hypoxia by adequate resuscitation has been shown to reduce mortality in sepsis [17]. Pharmacologic support is used to optimize cardiac output and mechanical ventilation is used to maintain oxygenation. Additional therapies, including renal replacement therapy, can be used to take over failing organ function. Although mortality rate of MODS is high, supportive measures can be tapered off in survivors, including renal replacement therapy, indicating that organ dysfunction is reversible. A number of treatment options may be possible to enhance mitochondrial function during this window of opportunity.

Regulating Cellular Substrate

In recent years, some evidence has emerged that may indicate that efforts to improve bio–energetic failure by regulating cellular substrate supply are beneficial in the critically ill. Hyperglycemia is a common finding in critically ill patients, as a result of stress–induced insulin resistance and accelerated glucose production. Intensive insulin treatment aiming at maintaining normoglycemia was shown to reduce mortality in patients on a surgical intensive care unit, as well as reduce inflammation and the occurrence of MODS [18, 19]. The protective effect of normoglycemia may occur via maintenance of mitochondrial integrity. In a post–mortem study, liver mitochondria from patients who were assigned intensive insulin therapy showed less morphological abnormalities when compared to patients assigned conventional therapy, which correlated with a higher activity of respiratory chain complexes I and IV [20]. Of interest, mitochondrial structure and function in skeletal muscle were not affected by intensive insulin therapy. In liver, glucose uptake is mediated by GLUT–2, independent of insulin, whereas insulin is required for GLUT–4 mediated glucose uptake in skeletal muscle [21]. Sepsis–induced resistance to insulin thus allowed for uptake of glucose in hepatocytes but not myocytes. Therefore, the mitochondrial abnormalities in sepsis patients may have resulted from direct toxic effects of high blood glucose levels [20, 22].

Another cellular substrate which has been studied to improve mitochondrial function in sepsis is succinate. Unlike complex I, complex II is relatively preserved during sepsis. Succinate is a component of the citric acid cycle and specifically donates electrons to complex II of the electron transfer chain, bypassing complex I. In an ex vivo rat model of sepsis, the addition of succinate was found to increase mitochondrial oxygen consumption [23]. The clinical significance of this strategy remains to be explored.

The effect of supplementing several amino acids have been studied in the critically ill too. Arginin, an NO–donor, increases protein synthesis and improves immunologic host defense during sepsis. Arginin–enriched enteral feeding formulae have been found to decrease the occurrence of multiple organ failure in critically ill trauma patients [24], and to reduce mortality of septic patients [25]. Glutamine is a precursor of the anti–oxidans gluthathion. Improving the balance between overproduction of reactive oxygen species and depletion of antioxidants during sepsis, may prevent the generation of peroxynitrite within the mitochondria, thereby restoring mitochondrial function. In accordance, in a model of sepsis, glutamine increased mitochondrial oxygen consumption, as exemplified by an increase in ATP-synthesis [26]. In septic patients, a supplement containing glutamine dipeptides, ant oxidative vitamins and trace elements, resulted in faster recovery from organ dysfunction compared to control patients [27], possibly by restoring low plasma levels of glutathione [28].

Induction of a Hypometabolic State

Instead of enhancing oxygen delivery, an alternative approach may be to reduce energy consumption. The regulated induction of a hypo–metabolic state, analogous to hibernation, may be beneficial in the imbalance between oxygen delivery and demand, thereby protecting the cells from severe bio-energetic failure and a critical fall in ATP.

Induced Hypothermia as a Novel Therapeutic Therapy in Multiple Organ Failure

Application of Hypothermia in Hypoxia-Induced Organ Damage

Induced hypothermia by external cooling is a well–known beneficial preventive strategy in conditions causing tissue injury, such as cardiothoracic surgery and organ transplantation. In addition, cooling the body to 32–34°C ameliorates neurological outcome when applied in patients that have suffered a cardiac arrest [29]. Other causes of hypoxic brain damage may also benefit from hypothermia, including stroke, traumatic brain injury and spinal cord injury [30]. Studies in experimental settings indicate that hypothermia may be protective in other organs suffering from of hypoxia-induced injury [31-33]. The beneficial effect is thought to occur via preservation of energy metabolism and reduction of the inflammatory response. Hypothermia reduces metabolism by 7% per grade, with reduction of ATP formation and reduction of cellular oxygen and cerebral glucose requirements. NO, which is produced in excess during sepsis, competes with oxygen in binding to cytochrome c oxidase in the mitochondrial membrane, thereby blocking the electron transport chain and resulting in overproduction of free oxygen radicals. Mild to moderate hypothermia prevents the production of superoxide and subsequent formation of reactive oxygen and nitrogen species during ischemia [34]. Another protective mechanism may include the prevention of apoptosis, which may also be linked to mitochondrial function [30]. The recovery of multiple organ failure has been found to be associated with improvement in mitochondrial respiration in survivors of septic shock [14]. As discussed above, both a lack of oxygen as well as an inability to utilize oxygen is likely to contribute to organ failure during critical illness. We hypothesize that hypoxic–induced organ damage in critically ill patients may benefit from induced hypothermia, by preservation of residual mitochondrial function or faster mitochondrial recovery after inflammation has resolved.

Acute Lung Injury

In 30–60% of critically ill patients, acute lung injury occurs in the course of an exaggerated inflammatory host response during MODS, most often mandating mechanical ventilation [35]. The mechanisms that contribute to acute lung injury involve inflammatory processes as well as mechanical processes due to overstretching of alveoli and repeated opening and collapsing of small airways. Reducing mechanical stress is a very beneficial strategy in these patients: the use of lower tidal volumes (preventing volutrauma) during mechanical ventilation has been found to reduce pulmonary damage in critically ill patients [36]. Besides too large tidal volumes, too frequent repetitive strain of respiratory cycles also may cause lung injury. Lowering of respiratory frequency (preventing tachytrauma), attenuated lung damage in experimental models [37]. However, the use of low tidal volumes and lower respiratory rates is limited by the fact that the resulting low minute ventilation results in high levels of arterial PCO_2 and concomitant severe respiratory acidosis.

Effects of Hypothermia on Acute Lung Injury

In animal models, induced hypothermia has been found to attenuate lung injury via reduction of neutrophil–mediated inflammation [38, 39]. Hypothermia may also exert protective effects by its effect on metabolism. Reduced CO_2–production and O_2–demand may allow for lower minute ventilation. Indeed, it was found that hypothermia allowed for mechanical ventilation using a low respiratory rate, thereby attenuating lung injury in a rat model, a strategy which was termed "lung rest" [40].

The clinical significance of hypothermia during acute lung injury has been shown in an earlier study. In moribund patients with severe acute lung injury, hypothermia applied as a last resort was found to reduce mortality [41]. However, progress has been made since this trial, and treatment of the critically ill patients has changed considerably. Whether the beneficial effects of hypothermia can be reproduced in less severely ill patients with acute lung injury, awaits exploration.

Effects of Hypothermia on Acute Kidney Injury

Acute kidney injury shows a striking similarity to the inflammatory reaction observed in acute lung injury. An exaggerated inflammatory response, including the induction of cytokines and the initiation of coagulation, contributes to acute kidney injury [42]. Another defining feature is the damage to the microvascular endothelium and epithelium leading to altered blood flow and oxygen extraction, as well as an increased permeability to proteins and solutes. Notably, acute lung injury may induce acute kidney injury. Deterioration of kidney function in the course of acute lung injury, carries a poor prognosis [43]. Data on the effect of hypothermia on hypoxia–induced kidney injury are limited to ischemia–induced kidney injury associated with the use of cardiopulmonary bypass. In series of patients, hypothermia applied during cardiopulmonary bypass for aortic surgery has been reported to protect against renal failure [44, 45]. It can be hypothesized that the protective effect of hypothermia found in models of acute lung injury, also apply to acute kidney injury.

Effects of Hypothermia on Gut Ischemia

In sepsis–induced multiple organ failure, microcirculatory abnormalities may depress gut barrier function and contribute to bacterial translocation. In critically ill patients, increased intestinal permeability was found to be predictive of the development of MODS [46]. In a model of intestinal ischemia–reperfusion injury, hypothermia reduced the amount of injury, which was related to both a reduction in neutrophil-infiltration as well as to a complete recovery of hepatic ATP synthesis [33]. The mechanism of this protective effect may have been inhibition of NO–mediated oxidative stress, as hypothermia attenuated NO–production and prevented depletion of gut glutathione [47]. Interestingly, hypothermia applied during gut ischemia, shifted cardiac substrate utilization from fatty acid oxidation to carbohydrate, as shown by an inhibition of carnitine palmitoyl transferase I activity [48]. The importance of mitochondrial dysfunction in this model was exemplified by the correlation between preservation of hepatic ATP levels and mortality [49].

Risk of Adverse Events

Induction of systemic hypothermia is feasible in patients. Adverse events, in particular infections and bleeding, are not increased during hypothermia when compared to normothermia in patients after a cardiac arrest [29]. Therefore, in experienced hands, hypothermia is a safe treatment in these patients. However, hypothermia has not been applied before in patients with acute lung injury or other organ failure occurring during critical illness. A number of issues need to be explored before clinical application. It can be hypothesized that hypothermia hampers adequate immune response during pneumonia or other bacterial infections, possibly leading to diminished clearance of bacteria. Indeed, a role for mediators that are produced during fever has been suggested for adequate host defense against bacteria [50]. Also, the effect of hypothermia on the risk of bleeding in the critically ill is unknown. Hypothermia results in prolonged bleeding time and diminished platelet aggregation. Although pulmonary bleeding is not a major feature of acute lung injury, hypothermia may increase the risk of bleeding from inflamed lung tissue with an altered morphology.

Suspended Animation as a Novel Therapeutic Therapy in Multiple Organ Failure

Another possible strategy that could be considered during critical illness is the reduction of cellular energy expenditure, alike hibernating animals when confronted with an environmental hypoxic insult. Hibernating mammals are thought to be tolerant to hypoxic conditions, by a regulated suppression of ATP–demand and ATP–supply to a new hypometabolic steady state [51]. Although humans do not hibernate naturally and have a limited tolerance to inadequate oxygenation, anecdotal reports on survival of prolonged episodes of profound hypothermia and circulatory arrest [52] suggest that the ability to switch to a hypometabolic state with lowered oxygen consumption may latently be present.

A hibernation–like state has been induced in non–hibernating animals with the use of hydrogen sulfide (H_2S) [53]. By competing with oxygen in binding to cytochrome c oxidase, H_2S can inhibit mitochondrial respiration, thereby reducing cellular oxygen consumption. Mice exposed to H_2S had a drop in core body temperature and a concomitant drop in metabolic rate, as measured by decreased O_2–consumption and CO_2–production. After cessation of H_2S exposure, the mice awoke, without displaying neurological or behavioral deficits. H_2S–induced suspended animation has been studied in several larger animal models, including sheep and pigs [54, 55]. These experiments have yielded conflicting results, which may have resulted from differences in experimental set up, including the use of different H_2S donor compounds as well as the use of anesthetic agents that may have influenced oxygen consumption. Conceivably, a difference in body mass may also contribute to these differences. Due to a large surface to mass ratio, small animals can rapidly reduce core body temperature, which may be difficult to induce in larger mammals and humans. However, in several experiments, the metabolic effect of H_2S occurred before core body temperature had dropped, suggesting that the suppressive effects of hibernation on metabolism are independent from the effects on body temperature [54, 56]. Also, thermal inertia of large mammals did not prevent the induction of profound hypothermia in former experiments [57]. Therefore, a suspended animation–like state may be feasible in large mammals and humans.

The effect of inducing a suspended animation–like state in disease models

H_2S has been found to protect against myocardial ischemia–reperfusion injury in non–hibernating doses [58-60], as well as in a dose that preserved mitochondrial structure and function compared to controls [61], via a vaso–relaxant effect [58], attenuation of inflammation [61] and reduction of apoptosis [59, 61]. In a model of trauma–induced acute lung injury, H_2S at high doses attenuated lung injury, by decreasing pro-inflammatory cytokines and upregulating anti–inflammatory cytokines [62]. In addition, an antioxidant effect of H_2S was observed.

A possible protective effect of H_2S on oxygen deprivation is of particular interest in critically ill patients suffering from severe lung injury, in which oxygenation can be barely maintained using potentially damaging high pressure mechanical ventilation. Of interest in this respect, are experiments with carbon monoxide, which acts as an oxygen reducer, similar to H_2S. Inhibition of cytochrome c oxidase by carbon monoxide can protect nematodes against severe hypoxia by inducing suspended animation [63]. This has led to the suggestion that suppressing oxygen demand before oxygen supply falls short, may protect the generation of reactive oxygen species and subsequent cell damage. In support of this hypothesis, it was shown that prior exposure of mice to H_2S increased survival during lethal hypoxia by a reduced oxygen demand [64].

Issues to Be Considered

The induction of a suspended animation–like state is still far from clinical application. When larger body masses require higher doses of H_2S to induce a suspended animation–like state, the therapeutic window may be too small, increasing the risk of toxicity. Toxicity has been extensively studied from the environmental viewpoint, reviewed in [65]. Dose–finding studies into hibernation–like states are mandated in appropriately–sized animal models. Of note, even in small doses, H_2S may exert toxic effects. In vitro, H_2S has been found to promote apoptotic cell death [66]. In models of sepsis, inhibition of endogenous H_2S was found to mediate or aggravate organ inflammation [67, 68]. In contrast, also anti–inflammatory effects have been found [62]. Furthermore, similar to the caveat described during hypothermia, adverse effects of inhibiting metabolism on host defense during infection need to be addressed.

Toxicity and flammability form practical hurdles for clinical implementation of H_2S. However, this objection has been overcome with other flammable gases such as oxygen, as well as other 'toxic' gases, such as NO. In addition, H_2S has the smell of rotten eggs. In a closed system of mechanical ventilation, exposure of patients and personnel to the odor may be limited. Lastly, corrosion of tubes and metal parts may shorten durability of the mechanical ventilator.

Conclusion

Mitochondrial dysfunction plays a role in critically ill patients with MODS. Preclinical evidence suggests that inducing a hypo–metabolic state limits organ injury by inhibition of the inflammatory response and by preservation of mitochondrial function, possibly via reduction of oxidative stress. Restoring the imbalance between oxygen demand and consumption may provide a novel therapeutic approach towards critically ill patients.

References

[1] Landry DW, Oliver JA. The pathogenesis of vasodilatory shock. *N. Engl. J. Med.* 2001;345:588-95.

[2] Balk RA. Pathogenesis and management of multiple organ dysfunction or failure in severe sepsis and septic shock. *Crit. Care Clin.* 2000;16:337-52, vii.

[3] Ince C, Sinaasappel M. Microcirculatory oxygenation and shunting in sepsis and shock. *Crit. Care Med.* 1999;27:1369-77.

[4] Boekstegers P, Weidenhofer S, Pilz G, Werdan K. Peripheral oxygen availability within skeletal muscle in sepsis and septic shock: comparison to limited infection and cardiogenic shock. *Infection.* 1991;19:317-23.

[5] VanderMeer TJ, Wang H, Fink MP. Endotoxemia causes ileal mucosal acidosis in the absence of mucosal hypoxia in a normodynamic porcine model of septic shock. *Crit. Care Med.* 1995;23:1217-26.

[6] Gotloib L, Shostak A, Galdi P, Jaichenko J, Fudin R. Loss of microvascular negative charges accompanied by interstitial edema in septic rats' heart. *Circ. Shock.* 1992;36:45-56.

[7] Welty-Wolf KE, Simonson SG, Huang YC, Fracica PJ, Patterson JW, Piantadosi CA. Ultrastructural changes in skeletal muscle mitochondria in gram-negative sepsis. *Shock.* 1996;5:378-84.

[8] Singer M. Mitochondrial function in sepsis: acute phase versus multiple organ failure. *Crit. Care Med.* 2007;35:S441-S448.

[9] Liaudet L, Soriano FG, Szabo C. Biology of nitric oxide signaling. *Crit. Care Med.* 2000;28:N37-N52.

[10] Brealey D, Karyampudi S, Jacques TS, et al. Mitochondrial dysfunction in a long-term rodent model of sepsis and organ failure. *Am. J. Physiol. Regul. Integr. Comp. Physiol.* 2004;286:R491-R497.

[11] Callahan LA, Supinski GS. Sepsis induces diaphragm electron transport chain dysfunction and protein depletion. *Am. J. Respir. Crit. Care Med.* 2005;172:861-8.

[12] Simonson SG, Welty-Wolf K, Huang YT, et al. Altered mitochondrial redox responses in gram negative septic shock in primates. *Circ. Shock.* 1994;43:34-43.

[13] Beltran B, Mathur A, Duchen MR, Erusalimsky JD, Moncada S. The effect of nitric oxide on cell respiration: A key to understanding its role in cell survival or death. *Proc. Natl. Acad. Sci.U. S. A.* 2000;97:14602-7.

[14] Brealey D, Brand M, Hargreaves I, et al. Association between mitochondrial dysfunction and severity and outcome of septic shock. *Lancet.* 2002;360:219-23.

[15] Fredriksson K, Hammarqvist F, Strigard K, et al. Derangements in mitochondrial metabolism in intercostal and leg muscle of critically ill patients with sepsis-induced multiple organ failure. *Am. J. Physiol. Endocrinol. Metab.* 2006;291:E1044-E1050.

[16] Singer M, De S, V, Vitale D, Jeffcoate W. Multiorgan failure is an adaptive, endocrine-mediated, metabolic response to overwhelming systemic inflammation. *Lancet.* 2004;364:545-8.

[17] Rivers E, Nguyen B, Havstad S, et al. Early goal-directed therapy in the treatment of severe sepsis and septic shock. *N. Engl. J. Med.* 2001;345:1368-77.

[18] Van den BG. Insulin therapy for the critically ill patient. *Clin. Cornerstone.* 2003;5:56-63.

[19] Van den BG, Wilmer A, Hermans G, et al. Intensive insulin therapy in the medical ICU. *N. Engl. J. Med.* 2006;354:449-61.

[20] Vanhorebeek I, De VR, Mesotten D, Wouters PJ, De Wolf-Peeters C, Van den BG. Protection of hepatocyte mitochondrial ultrastructure and function by strict blood glucose control with insulin in critically ill patients. *Lancet.* 2005;365:53-9.

[21] Tirone TA, Brunicardi FC. Overview of glucose regulation. *World J. Surg.* 2001;25:461-7.

[22] Vanhorebeek I, Langouche L, Van den BG. Glycemic and nonglycemic effects of insulin: how do they contribute to a better outcome of critical illness? *Curr. Opin. Crit. Care.* 2005;11:304-11.

[23] Protti A, Carre J, Frost MT, et al. Succinate recovers mitochondrial oxygen consumption in septic rat skeletal muscle. *Crit. Care Med.* 2007;35:2150-5.

[24] Moore FA, Moore EE, Kudsk KA, et al. Clinical benefits of an immune-enhancing diet for early postinjury enteral feeding. *J. Trauma.* 1994;37:607-15.

[25] Galban C, Montejo JC, Mesejo A, et al. An immune-enhancing enteral diet reduces mortality rate and episodes of bacteremia in septic intensive care unit patients. *Crit. Care Med.* 2000;28:643-8.

[26] Markley MA, Pierro A, Eaton S. Hepatocyte mitochondrial metabolism is inhibited in neonatal rat endotoxaemia: effects of glutamine. *Clin. Sci.* (Lond) 2002;102:337-44.

[27] Beale RJ, Sherry T, Lei K, et al. Early enteral supplementation with key pharmaconutrients improves Sequential Organ Failure Assessment score in critically ill patients with sepsis: outcome of a randomized, controlled, double-blind trial. *Crit. Care Med.* 2008;36:131-44.

[28] Luo M, Fernandez-Estivariz C, Jones DP, et al. Depletion of plasma antioxidants in surgical intensive care unit patients requiring parenteral feeding: effects of parenteral nutrition with or without alanyl-glutamine dipeptide supplementation. *Nutrition.* 2008;24:37-44.

[29] Mild therapeutic hypothermia to improve the neurologic outcome after cardiac arrest. *N. Engl. J. Med.* 2002;346:549-56.

[30] Polderman KH. Induced hypothermia and fever control for prevention and treatment of neurological injuries. *Lancet.* 2008;371:1955-69.

[31] Gotberg M, Olivecrona GK, Engblom H, et al. Rapid short-duration hypothermia with cold saline and endovascular cooling before reperfusion reduces microvascular obstruction and myocardial infarct size. *BMC Cardiovasc. Disord.* 2008;8:7.

[32] Shoji T, Omasa M, Nakamura T, et al. Mild hypothermia ameliorates lung ischemia reperfusion injury in an ex vivo rat lung model. *Eur. Surg. Res.* 2005;37:348-53.

[33] Stefanutti G, Pierro A, Parkinson EJ, Smith VV, Eaton S. Moderate hypothermia as a rescue therapy against intestinal ischemia and reperfusion injury in the rat. *Crit. Care Med.* 2008;36:1564-72.

[34] Small DL, Morley P, Buchan AM. Biology of ischemic cerebral cell death. *Prog. Cardiovasc. Dis.* 1999;42:185-207.

[35] MacCallum NS, Evans TW. Epidemiology of acute lung injury. *Curr. Opin. Crit. Care.* 2005;11:43-9.

[36] Amato MB, Barbas CS, Medeiros DM, et al. Effect of a protective-ventilation strategy on mortality in the acute respiratory distress syndrome. *N. Engl. J. Med.* 1998;338:347-54.

[37] Hotchkiss JR, Jr., Blanch L, Murias G, et al. Effects of decreased respiratory frequency on ventilator-induced lung injury. *Am. J. Respir. Crit. Care Med.* 2000;161:463-8.

[38] Chin JY, Koh Y, Kim MJ, et al. The effects of hypothermia on endotoxin-primed lung. *Anesth. Analg.* 2007;104:1171-8, tables.

[39] Chu SJ, Perng WC, Hung CM, Chang DM, Lin SH, Huang KL. Effects of various body temperatures after lipopolysaccharide-induced lung injury in rats. *Chest.* 2005;128:327-36.

[40] Hong SB, Koh Y, Lee IC, et al. Induced hypothermia as a new approach to lung rest for the acutely injured lung. *Crit. Care Med.* 2005;33:2049-55.

[41] Villar J, Slutsky AS. Effects of induced hypothermia in patients with septic adult respiratory distress syndrome. *Resuscitation.* 1993;26:183-92.

[42] Hassoun H, Grigoryev DN, Lie M, et al. Ischemic Acute Kidney Injury Induces a Distant Organ Functional and Genomic Response Distinguishable from Bilateral Nephrectomy. *Am. J. Physiol. Renal. Physiol.* 2007.

[43] Belperio JA, Keane MP, Lynch JP, III, Strieter RM. The role of cytokines during the pathogenesis of ventilator-associated and ventilator-induced lung injury. *Semin. Respir. Crit. Care Med.* 2006;27:350-64.

[44] Kouchoukos NT, Masetti P, Rokkas CK, Murphy SF. Hypothermic cardiopulmonary bypass and circulatory arrest for operations on the descending thoracic and thoracoabdominal aorta. *Ann. Thorac. Surg.* 2002;74:S1885-S1887.

[45] Kouchoukos NT, Masetti P, Murphy SF. Hypothermic cardiopulmonary bypass and circulatory arrest in the management of extensive thoracic and thoracoabdominal aortic aneurysms. *Semin. Thorac. Cardiovasc. Surg.* 2003;15:333-9.

[46] Doig CJ, Sutherland LR, Sandham JD, Fick GH, Verhoef M, Meddings JB. Increased intestinal permeability is associated with the development of multiple organ dysfunction syndrome in critically ill ICU patients. *Am. J. Respir. Crit. Care Med.* 1998;158:444-51.

[47] Stefanutti G, Pierro A, Vinardi S, Spitz L, Eaton S. Moderate hypothermia protects against systemic oxidative stress in a rat model of intestinal ischemia and reperfusion injury. *Shock.* 2005;24:159-64.

[48] Stefanutti G, Vejchapipat P, Williams SR, Pierro A, Eaton S. Heart energy metabolism after intestinal ischaemia and reperfusion. *J. Pediatr. Surg.* 2004;39:179-83.

[49] Vejchapipat P, Williams SR, Proctor E, Lauro V, Spitz L, Pierro A. Moderate hypothermia ameliorates liver energy failure after intestinal ischaemia-reperfusion in anaesthetised rats. *J. Pediatr. Surg.* 2001;36:269-75.

[50] Schroeder S, Bischoff J, Lehmann LE, et al. Endotoxin inhibits heat shock protein 70 (HSP70) expression in peripheral blood mononuclear cells of patients with severe sepsis. *Intensive Care Med.* 1999;25:52-7.

[51] Hochachka PW, Buck LT, Doll CJ, Land SC. Unifying theory of hypoxia tolerance: molecular/metabolic defense and rescue mechanisms for surviving oxygen lack. *Proc. Natl. Acad. Sci. U. S. A.* 1996;93:9493-8.

[52] Bernard S, Buist M, Monteiro O, Smith K. Induced hypothermia using large volume, ice-cold intravenous fluid in comatose survivors of out-of-hospital cardiac arrest: a preliminary report. *Resuscitation.* 2003;56:9-13.

[53] Roth MB, Nystul T. Buying time in suspended animation. *Sci. Am.* 2005;292:48-55.

[54] Haouzi P, Notet V, Chenuel B, et al. H2S induced hypometabolism in mice is missing in sedated sheep. *Respir. Physiol. Neurobiol.* 2008;160:109-15.

[55] Li J, Zhang G, Cai S, Redington AN. Effect of inhaled hydrogen sulfide on metabolic responses in anesthetized, paralyzed, and mechanically ventilated piglets. *Pediatr. Crit. Care Med.* 2008;9:110-2.

[56] Volpato GP, Searles R, Yu B, et al. Inhaled hydrogen sulfide: a rapidly reversible inhibitor of cardiac and metabolic function in the mouse. *Anesthesiology.* 2008;108:659-68.

[57] Behringer W, Safar P, Wu X, et al. Survival without brain damage after clinical death of 60-120 mins in dogs using suspended animation by profound hypothermia. *Crit. Care Med.* 2003;31:1523-31.

[58] Johansen D, Ytrehus K, Baxter GF. Exogenous hydrogen sulfide (H2S) protects against regional myocardial ischemia-reperfusion injury--Evidence for a role of K ATP channels. *Basic Res. Cardiol.* 2006;101:53-60.

[59] Sodha NR, Clements RT, Feng J, et al. The effects of therapeutic sulfide on myocardial apoptosis in response to ischemia-reperfusion injury. *Eur. J. Cardiothorac. Surg.* 2008;33:906-13.

[60] Zhu YZ, Wang ZJ, Ho P, et al. Hydrogen sulfide and its possible roles in myocardial ischemia in experimental rats. *J. Appl. Physiol.* 2007;102:261-8.

[61] Elrod JW, Calvert JW, Morrison J, et al. Hydrogen sulfide attenuates myocardial ischemia-reperfusion injury by preservation of mitochondrial function. *Proc. Natl. Acad. Sci. U. S. A.* 2007;104:15560-5.

[62] Esechie A, Kiss L, Olah G, et al. Protective effect of hydrogen sulfide in a murine model of acute lung injury induced by combined burn and smoke inhalation. *Clin. Sci.* (Lond) 2008;115:91-7.

[63] Nystul TG, Roth MB. Carbon monoxide-induced suspended animation protects against hypoxic damage in Caenorhabditis elegans. *Proc. Natl. Acad. Sci. U. S. A.* 2004;101:9133-6.

[64] Blackstone E, Roth MB. Suspended animation-like state protects mice from lethal hypoxia. *Shock.* 2007;27:370-2.

[65] Beauchamp RO, Jr., Bus JS, Popp JA, Boreiko CJ, Andjelkovich DA. A critical review of the literature on hydrogen sulfide toxicity. *Crit. Rev. Toxicol.* 1984;13:25-97.

[66] Baskar R, Li L, Moore PK. Hydrogen sulfide-induces DNA damage and changes in apoptotic gene expression in human lung fibroblast cells. *FASEB J.* 2007;21:247-55.

[67] Collin M, Anuar FB, Murch O, Bhatia M, Moore PK, Thiemermann C. Inhibition of endogenous hydrogen sulfide formation reduces the organ injury caused by endotoxemia. *Br. J. Pharmacol.* 2005;146:498-505.

[68] Zhang H, Hegde A, Ng SW, Adhikari S, Moochhala SM, Bhatia M. Hydrogen sulfide up-regulates substance P in polymicrobial sepsis-associated lung injury. *J. Immunol.* 2007;179:4153-60.

In: Glycolysis: Regulation, Processes and Diseases
Editor: Paul N. Lithaw
ISBN: 978-1-60741-103-1
© 2009 Nova Science Publishers, Inc.

Short Communication

The Anti-Ageing Effect of Enhanced Glycolysis; Another Role of the Warburg Effect

Hiroshi Kondoh * ***and Takeshi Maruyama***
Department of Geriatric Medicine, Graduate School of Medicine,
Kyoto University, 54 Kawahara-cho, Shogoin, Sakyo-ku, Kyoto, 606-8507, Japan

Abstract

Enhanced glycolysis is observed in most of cancerous cells and tissues, called as the Warburg effect. The clinical significance of the Warburg effect has been well established, while it is not completely clarified why and when cancer cells start to display and acquire such a characteristic metabolic property. Especially cancerous cells maintain enhanced glycolysis in tissue culture under standard condition (20% oxygen), which can not be explained by the cellular adapataion to hypoxic condtion via transcriptional factor HIF-1 (Hypoxia inducible factor-1) activation. Recent findings on senescent and cancer research discovered the unexpected role of the Warburg effect in protecting cells from oxidative damage. These anti-ageing effect of the Warburg effect can be a clue to understand pathophysiological impact of such metabolic shift in tumorigenesis.

1. Multi-Step Carcinogenesis and the Warburg Effect

It is well known that tumorigenesis involves a multi-step process associated with a series of genetic and epigenetic alterations(Wu and Pandolfi, 2001). The sequential application of carcinogens, for example, 7,12-dimethyl-benzan-thracene (DMBA), followed by the

* To whom correspondance should be addressed: hkondoh@kuhp.kyoto-u.ac.jp; Tel 0081(0) 75-751-3465; Fax 0081(0) 75-771-9784

treatment with phorbol esters (TPA), can result in skin tumors of animal model. In these experimental models, each step is called initiation, promotion, and progression, respectively, supporting the notion that cancer progression includes multi-step events.

From the pathological and clinical view point, the process of tumorgenesis *in vivo* can be divided into three or more stages. That is, immortalization, transformation, and metastasis(Braithwaite and Rabbitts, 1999). Among others, the acquisition of indefinite proliferative potentials, called cellular immortalization, would be involved in the very early stage *in vivo*. Then, to grow more aggresively and three dimentionally *in vivo*, these immortalized cells should be transformed to be anchorage-independent, resistant to contact inhibition, angiogenic etc. In the end stage, they easily detach from each other, more easily attach to and degradade matrix components, invade and migrate to other tissues (metastasis) (figure 1).

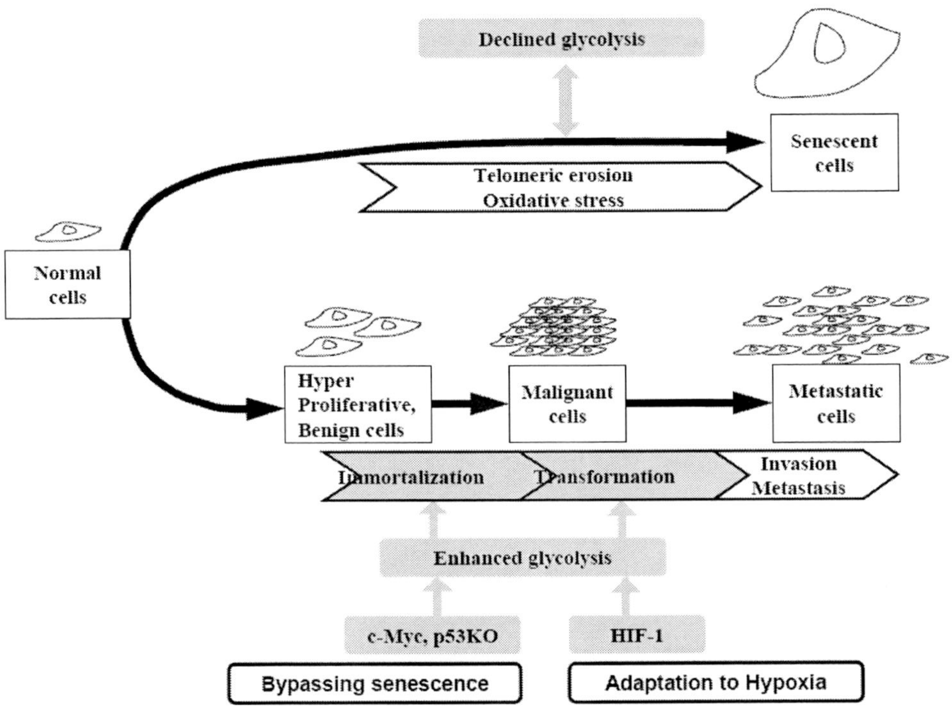

Figure 1. Modulation of glycolysis during multi-step oncogenesis and senescence. During multi-step tumorigenesis, enhanced glycolysis is required both in the step of immortalization and transformation. On the other hand, during senescent process, glycolysis declines. See text for the detail.

All these properties are also distinct hallmark of cancerous cells, compared with their normal counterpart(Hanahan and Weinberg, 2000). Another candiadate for inclusion in this list would be enhanced glycolysis, noted by Otto Warburg over seven decades ago. He first reported that cancerous tissues or cells display increased glycolysis by an unknown mechanism. A high glycolytic rate, even under high oxygen conditions, is referred to as the Warburg effect. This property is well utilized in clinical practice for the detection of metastatic tumor mass by positron-emission scanning of 2-[^{18}F]fluoro-2-deoxy-D-glucose. Thus there is no dispute about its clinical significance, while there has been a controversy when cancer cells become highly glycolytic *in vivo*. In other words, one big rising question is in which stage of multi-step oncogenesis the enhancement of glycolysis would be involved.

2. Enhanced Glycolysis Is Required During Transformation to Adapt Hypoxic Condition

It was widely assumed that cancer cells maintain upregulated glycolytic metabolism to adapt to the hypoxic condition *in vivo*, as solid aggressive tumors overgrow the blood supply of the feeding vasculatures(Dang and Semenza, 1999). Alternatively, it has been proposed that the increased glucose flux might improve efficiency of glucose utilisation in a microenvironment in which glucose is limited. In such a context, the glycolytic response represents a successful metabolic adaptation of cancer cells *in vivo*.

The concomitant induction of angiogenesis and enhancement of glycolysis with cell proliferation is mediated partly by activating hypoxia-inducible transcriptional factor (HIF-1)(Semenza, 1998). Hypoxia increases HIF-1α levels in most cell types and HIF-1 mediates adaptative responses to changes in tissue oxygenation. Thus, HIF-1 can directly upregulate expression of a set of genes involved in both local and global reaction to hypoxia, including angiogenesis, erythropoiesis, breathing and most of the glycolytic enzymes: hexokinase (HK1, HK2), Autocrine Motility Factor/Glucose-6-Phosphate Isomerase (AMF/GPI), enolase (ENO1), glucose transporter (GLUT1), glyceraldehyde-3-P dehydrogenase (GADPH), lactate dehydrogenase (LDH-A), 6-phosphofructo-2-kinase/fructose-2,6-bisphosphatase (PFKBF3), phosphofructokinase liver form (PFK-L), phosphoglycerate kinase (PGK1), pyruvate kinase muscle form (PK-M), triose phosphate isomerase (TPI). Interestingly, both HIF-1 and glycolytic enzymes are overexpressed in many tumors and cancer cells. Moreover, ectopic expression of glycolytic enzyme LDH can transform culture cells, while the inactivation of LDH ablate the transformation ability in cancer cells(Shim et al., 1997). Altogether, these data support a functional link between enhanced glycolysis and cellular adaptation during tumor formation and expansion. In conclusion, the transformation during multi-step oncogenesis would require the enhancement of glycolysis (figure 1).

However, the Warburg effect can not be explaind sololy by cellular adaptation to hypoxic condition for several following reasons.

1 Cellular adaptation model does not explain the constitutive metabolic change that maintains high glycolytic rates in cultured cancer cells even under 20% oxygen *in vitro(Gatenby and Gillies, 2004)*.
2 Ectopic expression of HIF-1 in culture cells induces cell cycle arrest, which argue if glycolysis in cancer cells is regulated simply by HIF-1 or hypoxia.
3 Indeed, recent works implicated that glycolysis is regulated in a cellular context dependent manner and by multiple genetic factors; oncogenes (ras, c-Myc), tumor suppressor genes (p53), signaling kinases (AMP kinase, Akt kinase, Pak1 kinase), and so on (figure 2).

These facts suggest that there would be a possible other mechanisms to increase glycolysis during multi-step oncogenesis other than adaptation to hypoxia.

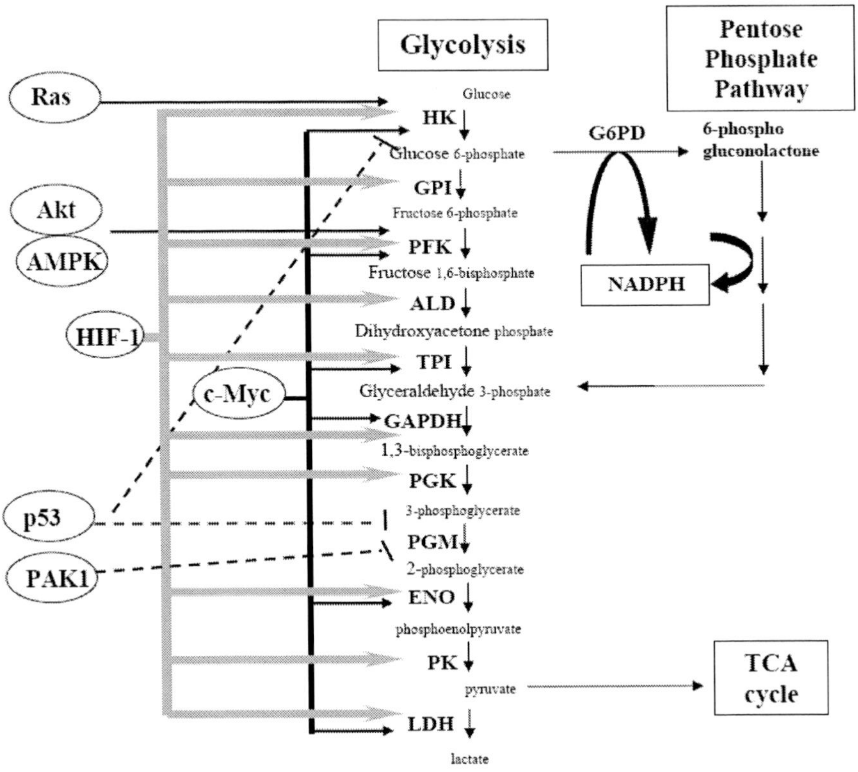

Figure 2. Regulatory pathway for glycolysis. Glycolytic pathway could be regulated by transcriptional factors (HIF-1, c-Myc), signaling kinases (Akt kinase, AMP kinase, Pak1 kinase), oncogenes (ras), tumor suppressors (p53), and others. Pentose Phosphate Pathway, a branching pathway from glycolytic pathway is crucial for NADPH production.

3. Enhanced Glycolysis Is Required Also in Immortalization

Recent studies implicates that the Warburg effect would be involved also in early stage of multi-step oncogenesis, that is, immortalization.

First, several immortalizing clones have been proved to function also as a booster for glycolysis.

For example oncogene c-Myc, an immortalizing clone for human primary epithelial cells and fibroblasts via telomerase activation(Wang et al., 1998), could also enhance glycolysis through transcriptional upregulation of several glycolytic enzymes; HK, PFK, TPI, GAPDH, ENO, and LDH(Kim et al., 2004). Inactivation of tumor suppressor gene p53 increases glycolytic flux *in vitro* and *in vivo*. p53 is most frequently mutated in various type of cancers, and functions as a transcriptional factor to induce cell cycle arrest, apoptosis, etc. Inactivation of p53 nicely immortalizes primary cells *in vitro*, implicating that p53 can also affect senescence process. p53 knock-out mice are cancer prone, possibly due to the senescence-inducing effect of p53. Partial activation of p53 induces premature organismal ageing *in vivo(Tyner et al., 2002)*.

Secondly, it was recently found that glycolytic enzymes can modulate cellular life span of MEFs(Kondoh et al., 2005b). In a senescence bypassing screening in MEFs using a retroviral cDNA library, the glycolytic enzyme phosphoglycerate mutase (PGM) was isolated as an immortalizing clone. PGM converts 3-phosphoglycerate to 2-phosphoglycerate during glycolysis. Analysis of the impact of other glycolytic enzymes over senescence in MEFs showed that glucophosphate isomerase (GPI) could also drive immortalization of MEFs. Ectopic expression of PGM or GPI increases glycolytic flux and extends the life span of primary MEFs. Conversely, knockdown of PGM or GPI via specific siRNA induces premature senescence. Noteworthy, several groups found that the glycolytic flux declines during senescence both in murine and human fibroblasts(Kondoh et al., 2005a).

Taken together, it is possible that glycolysis could be enhanced in the earlier step of tumorigenesis before the step of cellular transformation (figure 1).

4. Senescence-Bypassing Effect of Glycolysis via the Reduction of Oxidative Damage

Most somatic cells have a limited replicative capacity under standard tissue culture conditions and suffer a permanent cell cycle arrest, called replicative senescence(Wright and Shay, 2002). Replicative senescence is induced by telomere erosion upon reaching replicative exhaustion, which can be bypassed by the ectopic expression of telomerase in human fibroblasts.

Several stress can also induce premature senescence in a telomere-independent manner, called stress-induced senescence (SIS)(Sherr and DePinho, 2000). Several lines of evidence suggest that oxidative stress has the causal effects on premature senescence in a telomere-independent manner.

1 Mild oxidative stress (for example, treatment with low concentrations of hydrogen peroxide) is enough to induce senescence in primary cells.
2 During senescence process, the accumulation of oxidative damage in the cells is observed. Moreover most of immortalized cells are more resistant to the deleterious effects of oxidative damage than primary cells.
3 The reduction of oxidative stress by several means enables cells partly to bypass senescence; the ectopic expression of superoxide dismutase enzyme (SOD) as radical scavenger, tissue culture under hypoxic condition, the addition of radical scavenger (e.g., N-acetylcysteine) in the culture medium, and so on.

Thus, it is clearly established that these anti-oxidant scavengers are essential for proliferation of immortal cells, while little is known so far on the specific regulation operating in cancer cells.

5. Glycolysis as Radical Scavenger

As we discussed earlier, now it is known that the enhanced glycolysis can have the great impact on cellular senescence. But why and how such a metabolic shift can affect cellular lifespan? One possible explanation is that enhanced glycolysis can modulate oxidative stress, which can be a crucial trigger for cellular senescence. Indeed, MEFs immortalized by PGM or GPI suffer less oxidative damage than control cells as estimated by cytosolic ROS (Reactive Oxygen Species) staining, or quantification of 8-hydroxydeoxyguanosine (8-OHdG), a hallmark of oxidative DNA damage lesions.

Then how enhanced glycolysis can protect cells from oxidative damage? One clue is the common metabolic feature shared between several immortalized cells; cancer, immortalized primary, and ES cells(Kondoh et al., 2005a). They display enhanced glycolysis with decreased mitochondrial respiration. Thus, increased glycolysis during immortalization could be accompanied by the downregulation of mitochondrial function by unknown mechanism and would results in less intirinsic ROS production, as mitochondria is a major generator of ROS in the cells.

Second possible mechanism is that increased glycolysis could work as or activate radical scavenger. Interestingly, the anti-oxidant function of some scavengers (such as GSH or TRX) is closely coupled to the NADPH/NADP balance. Most of the NADPH/NADP is produced through the pentose phosphate pathway (PPP), a branching metabolic pathway derived from the glycolytic pathway. Enhanced glycolysis might activate PPP, and increase the level of NAPDH as by-products (figure 2).

Although this remains a hypothesis, several reports support it. The impact of glucose-6-phosphate dehydrogenase (G6PD) activity on cell proliferation is well established(Tian et al., 1999). G6PD catalyzes the rate-limiting step in the pentose phosphate pathway (PPP), which is responsible for the recycling of NADPH and maintenance of the redox balance as described above. G6PD-deficient human fibroblasts have a reduced lifespan that is attributable to oxidative stress and can be corrected by the ectopic expression of this enzyme(Ho et al., 2000). Both G6PD activity and the NADPH pool decline during continued culture passage, presumably as a consequence of the accumulation of oxidative damage. Importantly, ES cells ablated from G6PD expression are extremely sensitive to oxidative damage, showing massive apoptosis at low concentration of oxidants that are not lethal for wild type ES cells. It would, therefore, be worth exploring in the future whether enhanced glycolysis can then promote increased NADPH production via the PPP and exert its anti-senescence function.

6. Novel Regulatory Mechanism of Glycolysis

Recent adavance in understanding the regulatory mechanism of glycolysis identified tumor suppressor gene p53 as a key regulator of glycolysis *in vitro* and *in vivo*. Moreover, the ablation of p53 downregulates mitochondrial respiration by 30% *in vitro* and *in vivo*.

One possible interpretation would be that one of the major targets of p53 involved in senescence induction impacts on metabolic regulation, which render cells sensitive to oxidative stress. Such target is still unknown, but apparently p53 can modulate oxidative stress *in vivo*. Recent works suggest that knock-in mice for an extra copy of p53 acquire longevity associated with the resistance to oxidative damage(Matheu et al., 2007). Thus it is possible that the metabolic shift caused by p53 might modulate cellular senescence and organismal ageing through reduction of oxidative damage *in vivo and in vitro*.

p53 regulates both glycolysis and mitochondrial respiration through different targets. p53 can target glycokytic enzymes (HK and PGM) and modulate the overall flux of glycolysis in primary cells. On the other hand, p53 knockdown can decrease mitochondrial respiration(Kondoh et al., 2007). This is mailnly due to Cytochrome c Oxidase 2 (SCO2)(Matoba et al., 2006), a direct target by p53. SCO2 is critical regulating the cytochrome c oxidase complex, the major site of oxygen utilization. In this way, p53 could modulate both glycolysis and respiration in the concerted manner and would affect cellular lifespan. This fact suggests that stable alterations at the genetic or epigenetic levels, including inactivation of p53 may be the explanation for the enhanced glycolysis that cancer cells present *in vitro*.

In conclusion, glycolysis is essential both in the step of immortalization and transformation during multi-step tumorigenesis, as its enhancement can render cells resistant both to oxidative stress and hypoxic condition, respectively. Cancerous cells show concerted metabolic shift, including enhanced glycolysis with reduced mitochondrial respiration by poorly characterized mechanism. The regulation of glycolysis relevant to senescence process would be a key to improve and identify new anti-cancer therpy in the future.

References

Braithwaite, K.L., and Rabbitts, P.H. (1999). Multi-step evolution of lung cancer. *Semin. Cancer Biol. 9*, 255-265.

Dang, C.V., and Semenza, G.L. (1999). Oncogenic alterations of metabolism. *Trends in Biochemical Sciences. 24*, 68-72.

Gatenby, R.A., and Gillies, R.J. (2004). Why do cancers have high aerobic glycolysis? *Nat. Rev. Cancer. 4*, 891-899.

Hanahan, D., and Weinberg, R.A. (2000). The hallmarks of cancer. *Cell. 100*, 57-70.

Ho, H.Y., Cheng, M.L., Lu, F.J., Chou, Y.H., Stern, A., Liang, C.M., and Chiu, D.T. (2000). Enhanced oxidative stress and accelerated cellular senescence in glucose-6-phosphate dehydrogenase (G6PD)-deficient human fibroblasts. *Free Radic. Biol. Med. 29*, 156-169.

Kim, J.W., Zeller, K.I., Wang, Y., Jegga, A.G., Aronow, B.J., O'Donnell, K.A., and Dang, C.V. (2004). Evaluation of myc E-box phylogenetic footprints in glycolytic genes by chromatin immunoprecipitation assays. *Mol. Cell Biol. 24*, 5923-5936.

Kondoh, H., Lleonart, M.E., Gil, J., Beach, D., and Peters, G. (2005a). Glycolysis and cellular immortalization. *Drug Discovery Today. 2*, 263-267.

Kondoh, H., Lleonart, M.E., Gil, J., Wang, J., Degan, P., Peters, G., Martinez, D., Carnero, A., and Beach, D. (2005b). Glycolytic enzymes can modulate cellular life span. *Cancer. Res. 65*, 177-185.

Kondoh, H., Lleonart, M.E., Nakashima, Y., Yokode, M., Tanaka, M., Bernard, D., Gil, J., and Beach, D. (2007). A high glycolytic flux supports the proliferative potential of murine embryonic stem cells. *Antioxid. Redox. Signal. 9*, 293-299.

Matheu, A., Maraver, A., Klatt, P., Flores, I., Garcia-Cao, I., Borras, C., Flores, J.M., Vina, J., Blasco, M.A., and Serrano, M. (2007). Delayed ageing through damage protection by the Arf/p53 pathway. *Nature. 448*, 375-379.

Matoba, S., Kang, J.G., Patino, W.D., Wragg, A., Boehm, M., Gavrilova, O., Hurley, P.J., Bunz, F., and Hwang, P.M. (2006). p53 regulates mitochondrial respiration. *Science. 312*, 1650-1653.

Semenza, G.L. (1998). Hypoxia-inducible factor 1: master regulator of O2 homeostasis. *Curr. Opin. Genet. Dev. 8*, 588-594.

Sherr, C.J., and DePinho, R.A. (2000). Cellular senescence: mitotic clock or culture shock? *Cell. 102*, 407-410.

Shim, H., Dolde, C., Lewis, B.C., Wu, C.S., Dang, G., Jungmann, R.A., Dalla_Favera, R., and Dang, C.V. (1997). c-Myc transactivation of LDH-A: implications for tumor metabolism and growth. Proceedings of the National Academy of Sciences of the United States of America *94*, 6658-6663.

Tian, W.N., Braunstein, L.D., Apse, K., Pang, J., Rose, M., Tian, X., and Stanton, R.C. (1999). Importance of glucose-6-phosphate dehydrogenase activity in cell death. *Am. J. Physiol. 276*, C1121-1131.

Tyner, S.D., Venkatachalam, S., Choi, J., Jones, S., Ghebranious, N., Igelmann, H., Lu, X., Soron, G., Cooper, B., Brayton, C., *et al.* (2002). p53 mutant mice that display early ageing-associated phenotypes. *Nature. 415*, 45-53.

Wang, J., Xie, L.Y., Allan, S., Beach, D., and Hannon, G.J. (1998). Myc activates telomerase. *Genes Dev. 12*, 1769-1774.

Wright, W.E., and Shay, J.W. (2002). Historical claims and current interpretations of replicative aging. *Nat. Biotechnol. 20*, 682-688.

Wu, X., and Pandolfi, P.P. (2001). Mouse models for multistep tumorigenesis. *Trends Cell Biol. 11*, S2-9.

Index

B

D

E

F

H

I

J

K

N

O

P

Q

R

S